DISPLACEMENT

Derrida and After

A volume in the series
Theories of Contemporary Culture
Center for Twentieth Century Studies
University of Wisconsin–Milwaukee

General Editor, KATHLEEN WOODWARD

DISPLACEMENT

Derrida and After

EDITED WITH AN INTRODUCTION BY

Mark Krupnick

INDIANA UNIVERSITY PRESS

BLOOMINGTON

First Midland Book Edition 1987

Manufactured in the United States of America

Portions of "Jacques Derrida and the Heretic Hermeneutic" by
Susan Handelman first appeared in *The Slayers of Moses* by Susan A.
Handelman and are reprinted here by permission of the State University
of New York Press, copyright © 1982 State University of New York.

Library of Congress Cataloging in Publication Data
Main entry under title:

Displacement: Derrida and after.

(Theories of contemporary culture ; 5)
Includes bibliographical references.
Contents: Op writing / Gregory Ulmer—Staging,
Mont Blanc / Herman Rapaport—A trace of style /
Tom Conley—Derrida and the heretic hermeneutic /
Susan Handelman—[etc.]
I. Derrida, Jacques—Addresses, essay, lectures.
I. Krupnick, Mark, 1939-   . II. Series.
B2430.D484D57 1983   194   82-49301
cl ISBN 0-253-31803-3   pa ISBN 0-253-20455-0

2 3 4 5 6 91 90 89 88 87

FOR JEANNIE

CONTENTS

***

Acknowledgments

MANY PEOPLE ANSWERED my questions and otherwise contributed to my education in the course of the making of this book. I wish, first, to thank the contributors, several of whom were kind enough to read submissions I sent them and give me the benefit of their advice.

I am especially grateful to Herman Rapaport, who read nearly all of the papers in this collection and commented helpfully on the many questions I put to him. Jonathan Culler, Alexander Gelley, and Geoffrey Hartman were also particularly helpful. I am also grateful to G. Douglas Atkins, John Brenkman, Sydney Lévy, Roger McKeon, Françoise Meltzer, Stephen Melville, Andrew Parker, and Susan Shapiro for their advice.

Displacement is the fifth volume in the series on Theories of Contemporary Culture of the Center for Twentieth Century Studies of the University of Wisconsin–Milwaukee. Dean William Halloran and Associate Dean G. Micheal Riley of the College of Letters and Sciences of UW–M have supported the Center, and I am grateful to them for encouraging me in this project. I am grateful also to Kathleen Woodward, the director of the Center, who has kindly made available to me the Center's resources. These have included Naomi Galbreath's conscientious proofreading and assistance in styling the manuscript, and the able typing of Ginnie Schauble and Monica Verona.

Most of the work on this collection was done during the academic years 1980–82, when I was a Fellow of the Institute for Advanced Study of Religion at The Divinity School of the University of Chicago. For his generous hospitality, I am grateful to Professor Martin E. Marty, who was program coordinator at the Institute during my time there.

MARK KRUPNICK
February 1983

INTRODUCTION

***

Mark Krupnick

DERRIDA AND DISPLACEMENT

THIS COLLECTION TRACES A trajectory that runs from Sigmund Freud's *The Interpretation of Dreams* to Jacques Derrida's most recent writings. The separate essays are unified, first of all, by the pervasive presence of Derrida, the philosopher who has done most to displace the metaphysics of presence. Derrida has been far and away the most important European influence on the new American criticism. In this collection he is the cynosure of the essays directly about him (Handelman, Conley, Ulmer, Spivak), and he figures as an important influence on the other essays.

Displacement provides the organizing principle of this collection. The modern history of displacement begins with Freud's investigations of dreaming. Today, eighty years later, displacement has expanded its reference. No longer exclusively a technical term in psychoanalysis, it exists in a penumbra of meaning, summing up in its various connotations a decisive shift in humanity's understanding of itself. "Displacement" has become an indispensable term of the new post-structuralist theory, as much invoked, and with as little rigor, as "image" in the writings of the New Critics.

Freudian psychoanalysis (in Jacques Lacan's "return to Freud") also figures in this collection, and psychoanalysis and Derridean deconstruction mark the historical trajectory these essays assume. "Assume" rather then "specify," inasmuch as these essays are not studies in the history of ideas. Only Handelman and Spivak, in their very different approaches to Derrida, attempt anything like a history, and even they are closer in method to Derrida himself than to the kind of historiography Derrida attacks for its metaphysical presuppositions: "the history of meaning developing itself, producing itself, fulfilling itself. And doing so linearly . . . in a straight or circular line."[1]

Although displacement is not theoretically articulated in Derrida's writing, it is central to his de-centering mode of critique. For Derridean deconstruction proceeds by way of displacement, first reversing the terms of a philosophical opposition, that is, reversing a hierarchy or structure of domination, and then displacing or dislodging the system. Derrida speaks of "displacement" rather than "revolution" because of his sense "that the risk of metaphysical reappropriation is ineluctable," and that this reappropriation "happens very fast."[2]

A great deal is at stake in the newly important (dis-) place of displacement. In Derrida's writing, displacement almost always figures as an alternative to the Hegelian *Aufhebung*, the sublation by means of which contradictions are transcended. Under the new (post-Hegelian) dispensation, in the reign of difference (as opposed to identity), there will be no more grand claims, no more leapfrogging beyond stubborn conflicts to false reconciliations. Derrida is conspicuously modest in his conception of what deconstruction can accomplish. Deconstruction as he practices it allies itself with the voiceless, the marginal, the repressed, but it has no conviction that the old, bad (metaphysical) order can be transcended. The word is *déplacement* not *dépassement*. We may move things about, but we are not flattered into conceiving that we may "pass beyond."

What kind of essays, then, may the reader expect to find here? They are ambitious in scope, notably learned, and in several cases rather long. The day of the 5000-word New Critical explication—when all one needed was the text and mother wit—seems to be over.* These modernist critical texts are like modernist literary texts in transgressing conventional generic boundaries and in ignoring the conventional disciplinary limits of literary criticism. Broadly philosophical in their concern, these essays should be of interest to anyone concerned with questions of interpretation in psychoanalysis, theology, feminist writing, and political theory. They open up contemporary literary theory to our general concerns about sexual difference, the unconscious, and writing and representation.

These essays, with their large philosophical-cultural reference, are concerned, then, to illuminate theoretical issues rather than to expose the structure of individual works of art, and they commit any number of New Critical "heresies" in drawing on varieties of knowledge extrinsic to specific literary texts. But this new literature of theory is "interdisciplinary" in a new way. Continental theory has influenced our contemporary belief that it is impossible to think intelligently about literature without recourse to extraliterary knowledge. At the same time, this extraliterary knowledge (especially in the areas of linguistics, psychoanalysis, and philosophy) is not privileged in its relation to literature. It is itself problematized in the confrontation.

In general, these essays are about the cultural and ideological as well as the specifically belletristic implications of deconstruction. This matches

---

*One notes that the manifesto of the Yale critics, *Deconstruction and Criticism* (New York: Seabury, 1979), contains only five essays but takes up 256 book pages. The contributions of Harold Bloom, Paul de Man, Geoffrey Hartman, and Hillis Miller each run about forty pages, and Derrida's paper ("Living On: Border Lines") goes on for one hundred pages.

the expanding understanding of our advanced literary critics about just what "literature" is. The new subject of literary study includes society and culture and sexuality and the unconscious, all considered as *texts*. Literary study as now conceived extends beyond the poems of Donne and the novels of Henry James and the narrow, fetishizing New Critical notions of the work of art as urn or icon. *Displacement* is above all about writing itself, but it is also about women and the unconscious and Jewishness. For these are all exemplary instances of the marginal, and in the deconstructive analysis of texts the outsider returns to contest and demoralize authority. Deconstruction enables a return of the repressed, unsettling the law that gives priority to voice, patriarchy, rational consciousness, and the Greek-Christian logos. Deconstruction unsettles the idealisms that provide the ideological justifications for relations of power. Inevitably, then, a book about displacement that reflects the impact of Derrida on literary studies has a certain polemical aspect.

At the same time, *Displacement* showcases the efforts of some American critics and theorists in assimilating a new way of thinking about writing and textuality. One purpose of the book is to mediate this difficult new thinking, to provide the uninitiated with a (necessarily uncertain, displaced) foothold in crossing over from familiar ways of working with texts to unfamiliar modes of theorizing. Gradually all of Derrida's texts are being translated into English. The effect, already visible, is a revolution in methods of reading and analysis—thus far, in America, more in literary theory than in philosophy, the provenance of Derrida's own most influential readings. The deconstructions Derrida works on the texts of Rousseau and Hegel and Nietzsche we are learning to practice in our own sphere.

The 1970s were the first decade of the new French theory in American academic literary studies. These years appear in retrospect as the awkward age of post-structuralism in America. But American exponents of Derrida, Foucault, and Lacan are increasingly coming into their own, assimilating this body of theory without making it tame and "academic" in the bad sense. The present collection is a harbinger of this new American criticism. It suggests a new confidence, and even a capacity to displace (if not supersede) the French themselves, as we take what we need to help us along the way we need to go.

THE UBIQUITY OF DISPLACEMENT

Displacement traces a trajectory that runs from Freud to Derrida and, among our contemporaries, not Derrida alone. For "displacement" appears nearly everywhere in French theory and in the literary criticism it has influenced. If Hegelian language had not been placed under erasure by this new theory, we might say that "displacement" sums up the spirit of the present age. It is one of the indispensable words, like "text" itself, without which post-structuralism could not manage.

Derrida has given us an example in his own career of how a writer may successively abandon his working terms as he goes along. He has run through *différance*, *supplement*, *pharmakon*, *hymen*, *gram*, *spacing*, and several other equivalent terms: "certain marks," as he calls them, in an open chain. In the open chain of Derrida's successive deconstructive readings, these "marks" ceaselessly displace each other in a seemingly interminable procession. But "displacement" itself remains a constant and, in the rhetoric of contemporary literary theory, the one certitude you can count on amidst the dissemination that unsettles everything.

Examples come to hand all too readily. Thus the opening sentence of the editor's introduction to a recent book on post-structuralism:

> The name "post-structuralism" is useful in so far as it is an umbrella word, significantly defining itself only in terms of a temporal, spatial relationship to structuralism. This need not imply the organicist fiction of a development, for it involves, rather, a displacement.

"Displacement" in this context does not involve any epigenetic model of unfolding through time. "It is more a question of an interrogation of structuralism's methods and assumptions, of transforming structuralist concepts by turning one against another."[3] Displacement as "transformation" (as opposed to "translation"): this is a common sense of the word.

But if displacement is always with us in post-structuralist theory, it has no official status within it. It is no sacred word, unlike "tension" and "paradox" in the New Criticism, or "intertextuality" and "repetition" nowadays. And it is not likely ever to receive the sort of full-dress treatment that William Empson accorded his key terms in *The Structure of Complex Words* (1951). For displacement is not shrouded in ambiguities. It is ubiquitous without being complex and is used in myriad ways, not always rigorously, simply because critics take it for granted. What it stands for is an unquestioned presupposition of contemporary criticism.

Displacement became ubiquitous at that moment when an old poetics based on metaphor and symbol gave way to a new poetics that privileges metonymy. Roman Jakobson had argued that metonymy is as important as metaphor in human speech, but in literary-critical discourse it had not received much attention up to the present. And that may be because until recently literary theory has been a form of metaphysical speculation that privileges metaphor as the means of achieving the identity and self-sufficiency of the poem, its ontological presence. Metaphor, unlike metonymy, halts the attrition of time, making for an end to history. Metonymy, on the other hand, agrees with the contemporary impulse to reject any such spatializing of language. As such it has become an important element in what Paul de Man has called "the rhetoric of temporality." The pervasiveness of the word displacement implies our disillusioned rejection of the old dream of the filled and centered word, the incarnate word that was the model of an earlier criticism.

Geoffrey Hartman marks the transition to this newer criticism in some comments on Derrida's reading of Plato's *Phaedrus:*

> . . . can Derrida's analysis justify a massive displacement of interest from signified to signifier? More precisely, from the conceptualization that transforms signifier into signified to those unconceptualizable qualities of the signifier that keep it unsettled in form or meaning? Is the force of the written sign such that every attributed meaning pales before the originary and residual violence of a sound that cannot be fully inscribed because as sound it is already writing or incision. . . ?[4]

Displacement appears everywhere in criticism influenced by the idea of the diacritical character of language. But the word has a force, as Hartman implies, which will be lost on us if we think of displacement in terms of a genteel liberalism that makes for a kindly tolerance among the differences within the semiotic system. Derrida's "dissemination" or Gilles Deleuze's "nomad thought" suggest, on the contrary, a violence and a madness at work in language. Displacement is an exile from older certitudes of meaning and selfhood, a possibly permanent sojourn in the wilderness. Even when the political-juridical implications of Derrida and Deleuze are not in play, as in Paul de Man's studies in rhetoric, there remains a sense of this perilousness. As de Man writes: "Rhetoric radically suspends logic and opens up vertiginous possibilities of referential aberration."[5] One thinks of the title of Stephen Heath's book on Roland Barthes: *Vertige du déplacement*.

Tom Conley invokes displacement in his introduction to a special issue of the American journal *SubStance* (No. 31, 1982) devoted to *The Thing USA: Views of American Objects*. Conley writes that the articles in his special issue "look at recent sculpture, journalistic realism, titles attached to works of art, and above all they speak to an infinite dialogue written between objects on our own horizon and European modes of thinking that displace them."[6] The idea is that commonplace American artifacts that we no longer see freshly become unfamiliar, indeed "formidably strange," when they are removed from their habitual contexts and displaced into the very different context of French theory. The same idea in a new context— consider the difference between Heidegger's *immer schon* and Derrida's "always already"—is no longer the same idea. As Derrida has insisted all along, there is no translation without difference.

Conley's program of displacing American artifacts into a French theoretical context is similar to Shoshana Felman's project of displacement in the *Yale French Studies* special issue (No. 55–56, 1977) *Literature and Psychoanalysis*. In the Foreword, Felman describes her purpose of initiating a dialogue between French and American thinking, and "by means of this interaction, to de-center and to displace *both* the French and the American contexts and their ways of treating the topic (literature/psychoanalysis), so as to *put the topic in motion*. . . ."[7] The mutual displacement, that is, will affect not only the theme (American and French styles of criticism), but also the disciplinary relation of literature and psychoanalysis. No longer, as in the old days of "applied psychoanalysis," will one text (Freud's) serve as model-master-reference-explanation of another (the literary work of art); nor will literature be privileged in relation to psychoanalysis. The idea is to have them reciprocally displace each other, creating an oscillation. The ground slips beneath us: vertigo once more.

DISPLACEMENT IN THE TEXT OF FREUD

The contemporary interest in displacement as it bears on Freud is summed up by Paul Ricoeur in his paper of 1969 on "The Question of the Subject." In replying to "the challenge of semiology," Ricoeur argues that

> psychoanalysis has in no way eliminated consciousness and the ego; it has not replaced the subject, it has displaced it . . . consciousness and the ego still figure among the places and the roles which, taken together, constitute the human subject. The displacement of the problematic consists in the following: neither consciousness nor the ego is any longer in the position of principle or origin.[8]

Gayatri Spivak, in an essay written for this volume, takes a similar position. Spivak points out that what matters for deconstructive criticism is Freud's displacement of the subject. Not displacement as *Verschiebung*, Freud's word for the transfer of psychic energy from one idea to another in the process of dream-formation; not, that is, displacement as the complement of condensation, the other well-known operation of the dreamwork. Rather, what interests contemporary criticism is displacement as *Entstellung*, which is usually translated as "distortion" and is the word Freud uses to describe the dreamwork in general.

But we need to return to Freud himself and see just how he uses the word displacement and how our use of it differs from his. The most convenient summary of Freud's idea of displacement appears in *The Lan-*

guage of Psychoanalysis, by J. Laplanche and J.-B. Pontalis, a work that
presents the core concepts of psychoanalysis in dictionary form. The
entry for displacement begins with this definition:

> The fact that an idea's emphasis, interest or intensity is liable to be de-
> tached from it and to pass on to other ideas, which were originally of little
> intensity but which are related to the first idea by a chain of association.[9]

This definition needs some elaboration. In chapter six ("The Dream-
Work") of *The Interpretation of Dreams*, Freud describes displacement as
"nothing less than the essential portion of the dream-work."[10] It is the most
distinctive operation of the dream-work as of symptom-formation, because it involves the transfer of psychic energy from one idea to another.
This free movement of psychic energy, which can attach itself to, or
invest ("cathect"), one idea and now another, is a fundamental aspect of
the "primary process," which governs the unconscious system. In Freud
the displacement of energy is explained in terms of a "chain of associations" that traverse "associative pathways." The model is neurophysiolog-
ical, and Freud speaks of "energy" rather than of "desire," but it is not
hard to see how displacement might later be conceived, by Jakobson, as a
rhetorical operation (metonymy),[11] and, a few years later, by Lacan, as the
means by which desire is structured in the unconscious. This is how
Freudian displacement was itself displaced. The "associative pathways" of
Freud's *Project for a Scientific Psychology* (1895) reappear as the linguisticized
"signifying chains" in Lacan's important essay on "The agency of the
letter in the unconscious" (1957).[12]

Freud himself explained displacement in terms of the practical problems of clinical work and of his own self-analysis that led to its formulation. What, he wondered, caused the material of the manifest content of
the dream to be so different from the dream-thoughts as the latter were
revealed in the course of free association? Recounting the history of his
discovery, Freud notes that "it could be seen that the elements which
stand out as the principal components of the manifest content of the dream
are far from playing the same part in the dream-thoughts." Here Freud
invokes an idea (de-centering) that will become crucial for post-
structuralist theorizing: "The dream is, as it were, differently centered
from the dream-thoughts—its content has different elements as its central
point."[13]

He proceeds to gloss this idea using as examples several dreams he has
already discussed, including his own dream of the botanical monograph.
Freud might dream of a botanical monograph, but he learns from his self-
analysis that his dream-thoughts actually concerned professional anxieties
and ambitions. Does this mean that there is no connection between the

manifest content of the dream and the dream-thoughts, that the displacement of intensity from Freud's professional future to botany has no logic?
In asking about the random or determined character of this relation, we
ask about the continuity between (latent) thought and (manifest) scenario.

For Freud, relations are never random. In *On Dreams* (1901), a shortened
version of *The Interpretaion of Dreams*, he is discussing condensation in
relation to "overdetermination," and he observes: "A dream-element is, in
the strictest sense of the word, the 'representative' of all this disparate
material in the content of the dream."[14] The "strictest sense" would seem
to be parliamentary: every dream-element *stands for* many elements (con-
stituents?) in the dream-thoughts. But every dream-element is "representative" also in the sense that it signifies, or re-presents, elements in the
dream-thoughts. The relation between the botanical monograph in the
dream and the dreamer's wide-awake worries about his future is a *signifying* relation. It is not so clear that displacement in Derrida admits of
continuity and a relation of unequivocal signification.

THE GREAT DEBATE

By way of coming up to Derrida himself, let us cross over the ocean to
the scene of a clash that inaugurated the central debate of the entire past
decade in American academic literary criticism. The occasion of the clash
was the publication of a major work of American literary scholarship:
M. H. Abrams's *Natural Supernaturalism: Tradition and Revolution in Romantic Literature*. This work appeared in 1971, but it had been a long time
in the making and summarizes a rehabilitation of Romanticism, a rebuttal
to the modernist devaluation of nineteenth-century literature which had
been under way at least since the 1950s. When his book appeared, Ab-
ram's argument was not contested by old New Critical apologists for
Eliot, most of whom were dead or silent by this time. Rather, Abrams was
challenged by younger critics who had learned from Abrams himself and
who shared his desire to do justice to the Romantics. The critical difference was that the younger Romanticists had gone to school not only to
Abrams but also to Continental theory and had learned to think about
Romanticism and about literature differently.

In the Preface to *Natural Supernaturalism*, Abrams describes English and
German Romanticism as a "secularization of inherited theological ideas
and ways of thinking." But "secularization," he adds, did not mean rejection of religious ideas; rather it involved the "assimilation and reinterpretation" of these ideas in secular terms. The Romantics, according to
Abrams, "undertook . . . to save traditional concepts, schemes, and values
which had been based on the relation of the Creator to his creature and

creation, but to reformulate them within the prevailing two-term system of subject and object. . . ."

"Displacement" is Abrams's key word for this process of reinterpretation and reformulation: "Despite their displacement from a supernatural to a natural frame of reference . . . the ancient problems, terminology, and ways of thinking about human nature and history survived."[15] Romanticism was revolutionary, but it was also traditional; displacement in Abrams's sense makes for continuity. "Displacement" appears in a similar context in the body of his book as well. To understand Romantic philosophy and literature, Abrams writes, we must understand it as "displaced and reconstituted theology," a secularized-naturalized reformulation of the Christian story and Christian doctrines. And the process goes on: "we still live in what is essentially, although in derivative rather than direct manifestation, a Biblical culture," and we would do well to emulate Pierre Proudhon, who despite his militant atheism, "recognized his helplessness to escape religious formulas which, since they are woven into the fabric of our language, control the articulation of our thinking."[16]

The persistence of metaphysical-theological assumptions is an eminently Derridean theme, but this did not save Abrams from Hillis Miller's Derridean critique of *Natural Supernaturalism* in the journal *Diacritics* a year later.[17] The title of Miller's review is itself a displacement of Abrams's subtitle; Abrams's "Tradition and Revolution" becomes Miller's "Tradition and Difference." A Hegelian-idealist history of ideas is subjected to a Derridean deconstruction, and, again, "displacement" is invoked as a key term in the argument.

For Miller, Abrams's book is undermined by its metaphysical presuppositions, including the opposition of tradition and revolution. "Revolution" for Abrams is the Romantic ideal of a secular apocalypse, whether political or imaginative, which would be a humanist version, or "translation," of the Christian version of paradise regained. In Miller's view, "revolution" too much suggests the Hegelian model of alienation and the overcoming of alienation in a grand sublation. "Rather than the notion of revolution," Miller writes, "one needs the more enigmatic conception of repetition (repetition as displacement or decentering) to describe the effect of these writers on the culture. . . ."

Miller restates the idea a few pages later. He notes the important role in Romantic writing of the image of circle and center as a metaphor of the relation between God and the Creation. In terms of writing this would be the relation between an Urtext, a sacred Origin or transcendental signified outside the play of difference, and its re-presentations as copies. Miller writes that "In place of this image," of circle and center, "Nietzsche, Derrida, or Deleuze would put the notion of decentering or displacement, and the concept of a centerless repetition in which no element in the series is the commanding exemplar of which the others are copies."

In the course of this argument, Miller assets that there is no "essential continuity, no preservation of value and meaning, between the work and its 'source'" such as Abrams had proposed. An earlier text, Miller says, can never serve as unequivocal principle of explanation or meaning of the later text. In effect, Miller is rejecting the notion of displacement as guaranteeing continuity and at least putting into question in what sense displacement is a relation of signification.

DERRIDA'S POSITIONS

Positions, a collection of three interviews with Derrida, was published in France in 1972. This collection, which remains Derrida's clearest formulation of his critical project, appeared in an English translation in 1981. The most important of these interviews, itself entitled "Positions," was first published in successive issues of the French journal *Promesse* in 1971. This was the year of publication of Abrams's *Natural Supernaturalism*, but Abrams's thinking and Derrida's—about concepts, about history, about the task of criticism—are light years apart.

The term displacement does not have the special status of terms such as *différance, supplement, spacing,* and *dissémination* in Derrida's text. But the word does appear several times in "Positions," whenever Derrida is formulating his critical project, what he calls "a kind of *general strategy of deconstruction*." I shall cite these passages in the order of their appearance and try to show the resonances and implications of the term in each passage.

At the beginning of the interview Derrida is seeking to distinguish *différance* (neither a concept nor simply a word) from anything like it in Hegelian discourse. He acknowledges the stress on polysemy in the writings of Jean-Pierre Richard and Paul Ricoeur as an improvement on an older "monothematic" reading that always sought "the principal signified of the text, that is, its major referent." Derrida goes on:

> Nevertheless, polysemia, as such, is organized within the implicit horizon of a unitary resumption of meaning, that is, within the horizon of a dialectics . . . a teleological and totalizing dialectics that at a given moment, however far off, must permit the reassemblage of the totality of the text into the truth of its meaning, . . . annulling the open and productive displacement of the textual chain.[18]

The contrast here is between polysemy and Derrida's own concept/nonconcept of dissemination. Polysemy leads to a thematic criticism, a

totalization of meaning, a truth. Dissemination, on the other hand, is *Celbane's*
lawless and generative; its textual effects "can never be governed by a *idea of*
referent in the classical sense, that is, by a thing or by a transcendental *creation of the*
signified that would regulate its movement." Polysemy for Derrida seems *new myth.*
to suggest many meanings in one, and thus the ultimate possibility of a
transcendental resumption or recuperation of this multiplicity in a unity.
The idea of textual displacements, on the other hand and at its simplest,
would seem to be that there are no limits whatever to the inflections that a
word might acquire in its particular contexts.

Does this "open and productive displacement of the textual chain" establish new relations of signification? Perhaps so, but our usual language
for talking about meaning will not help us much. "Displacement" in Derrida's formulation connotes an unpredictable turbulence of signifiers
generated by "dissemination," a word that plays on the scattering of semes
like semen. But "scattering" is too mild. For Derrida, displacement involves a violent intervention: turbulence, irruption, explosion: "the deviance of meaning, its reflection-effect in writing, sets something off."

The second important mention of displacement occurs in the context of
the interviewer's questions about the bearing of Derrida's work on the
critique of ideologies. At a conference on this subject at Cluny, certain
conferees had dilated on Derrida's "thought," which was said—this was
April 1970—to be in "full evolution." Derrida, in the "Positions" interview, reacts sharply to this characterization. He describes himself as having "always been wary of 'thought.' No, it is a question of textual
displacements whose course, form, and necessity have nothing to do with
the 'evolution' of 'thought' or the teleology of a discourse."

"'Textual displacements' as against the "'evolution' of 'thought.'" Derrida unpacks this distinction himself, beginning by citing a statement of
his own from *Of Grammatology* (1967; English trans. 1976): "*In a certain
way, 'thought' means nothing.*"[19] Derrida goes on to say that "'thought' (is)
the illusory autonomy of a discourse or a consciousness whose hypostasis
is to be deconstructed. . . ." Derrida turns his argument also against
himself: "whatever will continue to be called thought, and which, for
example, will designate the deconstruction of logocentrism, means nothing, for in the last analysis it no longer derives from 'meaning.' Wherever
it operates, *'thought' means nothing.*"

A third important mention of displacement occurs in the context of a
discussion of history and historiography that I cited at the beginning of
this Introduction. Derrida is speaking of "the metaphysical character of
the concept of history" and the system of predicates to which it is linked
("teleology, eschatology, elevating and interiorizing accumulation of
meaning, a certain type of traditionality, a certain concept of continuity,

of truth, etc."). This metaphysical character, he goes on, "is not an accidental predicate which could be removed by a kind of local ablation,
without a general displacement of the organization. . . ."[20]

So, in relation to intellectual history, you cannot pass as M. H. Abrams
does from a metaphysical to a secular worldview without decisively altering meanings. Older forms of thinking and feeling *must* change with such a
displacement. Displacement itself takes on a new sense. It refers not to an
essentially conservative "reformulation" that has the effect of keeping the
best of the old while adapting to new circumstances. Instead, displacement now refers to a violent intervention intended to shake and demoralize that old order.

For Derridean deconstruction reverses a classical philosophical opposition (one of the violent hierarchies intrinsic to Western metaphysics), and
displaces or dislodges it, removes the ground. The drive of deconstruction
is precisely to *avoid* the reconciliation Abrams attributes to the High
Romantics. Although dedicated to teaching the whole of the Western
philosophical tradition, Derrida is more ambivalent than Abrams about
the idea of "the tradition" and the "certain type of traditionality" that goes
along with it.

In defining the deconstructive method, it is important to see that the
hierarchical opposition that is deconstructed is not thereby destroyed. For
example, the contrast between speech and writing is not eliminated;
rather, it is *displaced*, cut loose from its metaphysical grounding when it is
inverted, so that speech comes to be seen as a special case of an archi-
writing. With this displacement, speech can no longer function in the
same way. Derrida puts it in these terms. After the deconstructive reversal, "which brings low what was high," there is the displacement that
brings about "the irruptive emergence of a new 'concept,' a concept that
can no longer be, and never could be, included in the previous regime."[21]
This third item that is not a third term, this concept that is not a concept,
is outside the previous binary opposition and cannot be recuperated in a
dialectic. There can be no raising up, or sublation (Hegel's *Aufhebung*), of
the contraries.

NOVELTY OR REPETITION?

Political implications proliferate in Derrida's account of his project.
Deconstruction, he says, involves a "strategic" operation the purpose of
which is the effective transformation of the field. The "Positions" interview is full of such military metaphors, as Derrida draws on words like
intervention, force, surprise, violence. He says that the deconstructive
critique requires an "incision," a kind of surgical strike. This incision

"does not take place just anywhere." It must be made "according to lines *for 49*
of force" and sites of possible rupture where the discourse to be decon- *the system of*
structed is vulnerable. These are the nodes from which the given discur- *communication*
sive system takes its rise. Derrida describes these nodes as "holds" or
"levers" for a kind of Archimedean intervention. Like the Greek
philosopher, Derrida is able to move a very great weight by a very small
force. But, given Derrida's drive to ungrounding, one cannot imagine his
saying, as Archimedes is supposed to have said, "Give me a place to stand
on and I will move the earth."

Can a philosophic critique that has no ground to stand on, that exists in
order to remove all grounds, move the earth? In the "Positions" interview,
Derrida stresses efficacy, transformation, novelty. But this may be because his interviewer, a Marxist, seems primarily concerned with deconstruction as a philosophy of praxis. The interviewer, Jean-Louis
Houdebine, urges that Derrida acknowledge that *différance* is only another
word for "contradictions," and that deconstruction itself is a kind of Marxism. But although Derrida insists on the subversiveness of his method, he
spends most of the interview distinguishing his "positions" from the "position" (dialectical materialism) of the single-minded interviewer.

The question remains: what *is* the implication of deconstruction? Derrida is obviously concerned to avoid recuperation of any kind, as he demonstrates in successively abandoning what had appeared to be *princeps*
terms before they can become key words for a new thematic criticism.
And it is clear, in his commitment to the idea of difference (whatever
name he gives it), that he wants not to be assimilated to official philosophy. Still, his drift remains indeterminate, undecidable. Derrida wishes
to contest the official culture, but he does not confront it head-on or seek
to undermine it from below. Unlike Freud, a great precursor in deconstruction, Derrida has at no point in his career seemed to identify himself
with the great rebels of the Western tradition. One cannot imagine this
man of obliquities identifying with the epigraph from the *Aeneid* that
Freud placed at the beginning of *The Interpretation of Dreams:* "Flectere si
nequeo superos, Acheronta movebo" (If Heaven [or respectable culture]
remain unyielding, I shall move Hell).

A few years ago the central question about the new French theory, from
the point of view of its enemies, concerned what threat it represented to
Western civilization itself. To its liberal-humanist opponents, deconstruction was nasty, brutish, and (they hoped) short-lived. Since then the
debate has become more modulated; it is not as likely now to be conducted
in terms of such ultimacies. One question at present concerns how deconstruction can be opened up. What, we now wonder, are its implications
and its uses? This is a question about the implications of Derrida's think-

ing for a social-cultural criticism, for a new theory of history (including literary history), and for a new paradigm in literary studies.

The future of post-structuralist criticism will not depend exclusively on where Derrida goes from here. But he has been the single most powerful European influence on the new American criticism, and so his progress cannot be a matter of indifference. Lacking a crystal ball, we will need to study what he has already written, partly to see what there may be that qualifies his own sometime emphasis on the transgressive impulse in his thinking.

An important question concerns the relation between displacement and repetition. The work of deconstruction concludes with the reinscription in new chains of old terms. The old organization is shaken, and yet it remains. We think against metaphysical concepts but by means of words that cannot be cleansed of their metaphysical associations. "Reinscription" does not mean a simple repetition. Neither, however, does it guarantee a leap into novelty.

Writers influenced by Derrida have adapted the idea of originary difference to their own various ideological projects. But the will to power is less marked in Derrida's own writing than a certain will to relinquishment, a will to unfulfillment such as Jean-François Lyotard describes in his essay "Jewish Oedipus."[22] Derrida reminds us constantly how quickly the hierarchies of dual opposition reestablish themselves. Deconstruction offers no prospect of apocalyptic renovation. Instead, the deconstructive critic is committed to the necessity of interminable analysis. Metaphysical hierarchies and violent structures of domination will come back to haunt us. Deconstruction is steady work, needing to be repeated over and over.

Derrida's will to unfulfillment is suggested by the modesty of his claims, as also in his refusal of the vitalism of his philosophical contemporaries in France. One looks in vain in Derrida's writing for a politics of libido, such as Deleuze, Foucault, and Lyotard have advocated. As Susan Handelman observes in her essay in this collection, Derrida makes his home in Writing not in Nature.

Derrida has liberated others into a playful aestheticism, but his own thinking has a pessimistic and stoic side. His will to relinquishment goes along with a will to repeat—*Beyond the Pleasure Principle* is the Freud text he cites most. And what he repeats is the lesson of difference, which he struggles to keep from becoming a substance or fetish. Still, the effect is of the eternal recurrence of the same—as difference.

Working from inside the system he seeks to deconstruct, Derrida occupies his own *arête de mort* (ridge, or arris, of death), a kind of Promethean rock on which he repeats his exemplary acts of unweaving.[23] The

arête de mort is also an *arrêt de mort*, an undecidable term Derrida borrows
from Maurice Blanchot that may mean "death sentence" and/or "stay of
execution." If freedom may be a consequence of a pragmatic decision to
act as if one is free, so the decision to remain poised in indecision may also
have its practical consequences.

What is astonishing is the extreme scrupulousness of Derrida's self-
unweavings. There can be no simple inversions here, whereby the last
will be first and we get to line up our old foes (and lukewarm friends)
against the wall. In Derrida there are no final solutions—the last final
solution was enough. Instead, we are urged toward a scrupulously rigorous interminable analysis intended to fend off the inevitable recuperation
of those myths of presence and those structures of domination, which
always come back. The compulsion to repeat the demonstration of difference may be seen as Derrida's way of binding anxiety in the face of a
future which, as he wrote in the Exergue to *Of Grammatology*, "can only be
anticipated in the form of an absolute danger . . . as a sort of monstrosity."[24]

TO THE READER:

We are not going to introduce this book's individual essays in this
general introduction. For each part of the book there is an introduction
that establishes a context for reading each piece and points out connections
and common areas of meaning among the essays in that part. The in-
troductions also indicate how the individual essays bear on the general
topic of displacement.

Some recent expositions of post-structuralism have seemed to suggest,
either by their paucity or opacity of commentary, that this new body of
theory is more easily assimilated than in fact it is. The present collection,
in its effort of mediation, at least acknowledges the difficulty of coming to
terms with a theory displaced from a specific European intellectual context to the very different context of American academic literary studies.
The "implied reader" of these essays will probably at least have heard
some of the news from abroad and have dipped into Barthes or Derrida or
one or another of the important new quarterlies publishing advanced
literary theory. But the essays collected here do not assume we have all
personally lived through the literary-ideological vicissitudes of the avant-
garde French journal *Tel Quel* in the past two decades.

In the end, however, nobody can spare anyone else the labor of thinking
through the major primary texts, both philosophical and literary, of the
new critical theory, or the work involved in feeling one's way into the

papers collected here. If difficulty and difference are intrinsic to this new body of work, it is because, for the time being, criticism inhabits the wilderness. Literary theory, having discovered the principle of difference and displacement, remains too refractory to achieve anything like a neo-classical lucidity that would deny our actual condition in-between. There is no possibility in the near future of a critical consensus. And what kind of consensus could it be, founded on the ground(lessness) of difference, on the play of displacement?

When the consensus is achieved, it will not prove that post-structuralism has arrived, for displacement means *never arriving*. General acceptance will mean that this movement is all over, no longer in motion. Not *déplacé* but *dépassé*.

NOTES

1. Jacques Derrida, *Positions*, trans. Alan Bass (1972; Chicago: University of Chicago Press, 1981), p. 56.
2. Ibid., p. 58.
3. Robert Young, ed., *Untying the Text: A Post-Structuralist Reader* (Boston and London: Routledge and Kegan Paul, 1981), p. 1.
4. Geoffrey Hartman, *Saving the Text: Literature/Derrida/Philosophy* (Baltimore: Johns Hopkins University Press, 1981), pp. 119–20.
5. Paul de Man, *Allegories of Reading: Figural Language in Rousseau, Nietzsche, Rilke, and Proust* (New Haven: Yale University Press, 1979), p. 10.
6. Tom Conley, Introduction, *SubStance* 31 (1982).
7. Shoshana Felman, Foreword, *Literature and Psychoanalysis—the Question of Reading: Otherwise, Yale French Studies* 55–56 (1977).
8. Paul Ricoeur, "The Question of the Subject: The Challenge of Semiology," in *The Conflict of Interpretations: Essays in Hermeneutics* (Evanston: Northwestern University Press, 1974), p. 241.
9. Jean Laplanche and J.-B. Pontalis, *The Language of Psychoanalysis*, trans. Donald Nicholson-Smith (London: Hogarth Press, 1973), p. 121.
10. Sigmund Freud, *Standard Edition of the Complete Psychological Works*, trans. James Strachey, IV (London: Hogarth Press, 1964), p. 308.
11. Roman Jakobson, "Two Aspects of Language: Metaphor and Metonymy," in *European Literary Theory and Practice*, ed. Vernon Gras (New York: Delta, 1973), pp. 119–29.
12. Jacques Lacan, "The agency of the letter in the unconscious or reason since Freud," in *Écrits: A Selection*, trans. Alan Sheridan (New York: Norton, 1977), pp. 146–78.
13. Freud, *Standard Edition*, IV, p. 305.
14. Ibid., V, p. 652.
15. M. H. Abrams, *Natural Supernaturalism: Tradition and Revolution in Romantic Literature* (New York: Norton, 1971), pp. 12–13.
16. Ibid., pp. 65–66.
17. J. Hillis Miller, "Tradition and Difference," *Diacritics* 2 (Winter 1972): 6–13.

18. Derrida, *Positions*, p. 45.
19. Ibid., p. 49.
20. Ibid., p. 57.
21. Ibid., p. 42.
22. Jean-François Lyotard, "Jewish Oedipus," trans. Susan Hanson, *Genre* 10 (1977): 395–411.
23. Cf. Herman Rapaport's essay, "Staging: Mont Blanc," in this volume.
24. Derrida, *Of Grammatology*, trans. Gayatri Spivak (1967; Baltimore: Johns Hopkins University Press, 1976), p. 5.

Part One

***

Sensible Language

Introduction:
Sensible Language

"SENSIBLE LANGUAGE?" A JOKE, REALLY, for the criticism of Derrida and his American exponents is by no means "sensible" as that word would have been understood by earlier English-language critics like Edmund Wilson and Lionel Trilling or T. S. Eliot and F. R. Leavis. There are few echoes in Derridean writing of their older humanist positives, such as sincerity, tradition, clarity. The last of these, clarity, was once thought to be the rock on which French criticism was built. But now there are no more rocks and no established churches in criticism: not the liberal humanism of Matthew Arnold nor the Anglo-Catholic orthodoxy of Eliot nor the positivism of Raymond Picard (Roland Barthes's Establishment critic in the 1960s) and the older French literary scholarship. A too accommodating idea of reason and sense have been precisely the enemies Derridean criticism has been seeking to undermine. But what a crisis of critical language the revolt against *clarté* has brought about. Is there no way back to "sensible" language as Jane Austen, say, would have understood the word?

Not soon, if by "sensible" we refer to what all right-thinking persons can agree on, based on common sense and the evidence of the senses. However, "sensible" can mean something different, when it is opposed (in traditional idealist philosophy) to "intelligible," a word that suggests the transcendence of the senses. The "intelligible" in traditional Western metaphysics has often suggested the "spiritual," the volatilizing away of matter in its passage "beyond," to mind or idea. In the new French criticism, "sensible language" directs itself against this Platonizing translation into the ideal. So Derrida's language *is* "sensible" in attending to its own sensuous aspects, to its "aspect" or appearance rather than to the depths of a signified that underwrites appearance. Depths and heights are resisted in this new critical language for the sake of the surface. Without heights (Plato's eternal forms in the sky) and without depths (the underlying concealed meaning in conventional Freudian and Marxist hermeneutics), what are we left with?

The age of "sensible language" in criticism is the age of the (usually visual) pun: Gregory Ulmer's *oir* (*Moira, moiré, memoire,* even *theORIa,* slightly out of order); Herman Rapaport's seen-(ob)scene and the Freudian phantasm as a kind of pun; and Tom Conley's anagrams *(écart, trace, carte, cadre)*. Can these bones live? What can criticism make of such materials?

Gregory Ulmer's discussion of the analogy between Derrida's writing and Op Art takes off from Michel Foucault's observation that his generation of philosophers in France has been engaged in a search for new, transgressive models for thinking and writing to replace the traditional dialectical models. Ulmer's concern is with Derrida's new way of generating concepts, his displacement of the dialectical mode as it has dominated philosophy from Plato to Hegel and since. Inevitably this study of Derrida's mode of "concept"-formation (quotation marks because Derrida's "concepts" are not concepts in any commonly accepted sense) turns into a study of Derrida's way of writing—otherwise.

Not itself an exercise in deconstruction, Ulmer's essay is an exposition of Derrida's deconstruction of one of the founding metaphors of philosophy: *theoria*. The Greek word for "theory" combines *thea* (the visible aspect of things, thus linked to "theater") and *horao* (to look at something closely). Thus the term *theoria*, which points to the super-sensible world of Ideas, is itself composed of metaphors of sensible seeing. Derrida's critique of the privileged role of seeing in Western metaphysics is itself glossed, in Ulmer's essay, by examples from the visual arts (notably Op Art) that similarly raise questions about perception and optics.

Op Art makes objects tremble. So does "solicitation," the key term in Ulmer's title. For this Latinate word turns out to be another term for "displacement" or for "deconstruction" itself. By "solicitation," Derrida means to suggest rude energy, the force that upsets the static forms of classical structuralism. The word comes from the Latin *citare* (to put in motion) and *sollus* (in archaic Latin, the whole). Just so, Derrida's strategy is intended to shake up the structure, to cause it to tremble and vibrate. The certitude of "theory" gives way to the characteristic Derridean undecidability, the vertigo that has its analogue in the moiré effect (the flutter, oscillation, trembling) of Op Art.

Ulmer's essay seems an appropriate starting point for this collection because it clarifies the shaking-shuttling-oscillating intention of Derrida's textual displacements, and also because Ulmer draws on and quotes from such a wide variety of Derridean texts, across the whole span of his career. But the key texts for this essay are those writings of Derrida that bear directly on metaphor. For the second displacement that Ulmer takes up is within metaphor itself.

In classical rhetoric, we move vertically from the sensible to the spiritual; the materiality of the signifier is swallowed up in the ideality of the signified. Metaphor is thus the foundation of metaphysics, the sublation that reduces the signifier to a station on the way to one or another transcendental signified, that arrests the surface play of displacement. In his critique of representation, Derrida reverses the procedure. For him, ornament and style are not mere embellishment, vehicles of a pre-existing referent. Rather, ornament *generates* texts and concepts in a wholly abstract, nonrepresentational register. This is a writing that evades the strictures of voice and that seeks nonverbal models for thought-strategies capable of displacing conventional concept-formation. Thus the relevance, in Ulmer's view, of abstract design phenomena.

Ulmer's own ornament is *Moira*, the Greek word for "destiny" and the title of an important essay by Heidegger. There is no need to be any more solemn about *Moira* than Ulmer is himself. The word serves him not as a tragic idea or concept but more as a conceit, a way of opening things up; it is a kind of prop for thinking. It works, in Ulmer's text, somewhat in the way that Derrida says *différance* works for him: not as a "concept," nor simply as a "word," yet generating "conceptual effects" and "verbal concretions." *Moira* serves Ulmer as a prop for studying Derrida's notion that language is destiny. Thus the seemingly arbitrary conjuntion of *oir* words—*Moira, moiré, memoire, grimoire* (medieval book of spells, in *La carte postale*)—turns out to obey a certain (un-Aristotelian) logic: neither pure randomness nor pure determinism but something closer to Nietzsche's idea of history as "the iron hand of necessity shaking the dice box of chance."

Ulmer notes that Derrida was building a "rigorous, systematic discourse based on the equivocity of language" in the same years that the geometrical Constructivists of the School of Paris were building a rigorous, systematic visual art based on the equivocity of perception. In "Living On: Border Lines" (*Deconstruction and Criticism*, 1979), Derrida writes of "superimposing" one text on another. "Border Lines" is a running footnote-commentary that accompanies what is ostensibly a critique ("Living On") of Shelley's long poem *The Triumph of Life*. Just so, Op Art involves superimposition and induces a blurring known as the moiré effect. This is designing "otherwise," as Derrida's is writing otherwise.

Ulmer is arguing, then, that Derrida uses various optical effects from the visual arts to achieve a deconstruction of the optical effects sublated in the metaphors of theoretical discourse. Ulmer does not, however, insist on Op Art as an "influence" on Derrida. Rather, Op Art figures for Ulmer as an analogy for Derrida's nondialectical, nonrepresentational writing. To use Herman Rapaport's word, Op Art functions as a kind of "prop" in Ulmer's account of Derridean thinking. Indeed, to a certain extent Ulmer uses Derrida himself as a prop for his own post-Derridean thinking. For Ulmer's essay, offered modestly as an exposition of Derrida, does not simply elucidate the object of study. As with all of the essays in this collection, the writer takes his cue from Derrida but then goes his own way.

The concern with the visual has been part of the continuing critique of the traditional model of representation. In *Criticism in the Wilderness* (1981), Geoffrey Hartman has proposed a critique of the privileged status of the visual "image" in modernist poetics. Hartman wonders whether "Perceptibility—that all things can be made as perceptible as the eye suggests— may itself be the great classic phantasm, the mediterranean fantasy. . . ." The alternative to the traditional idea of the image, as residue of perception, is the idea of the phantasm. The phantasm as it destabilizes perception and displaces the seeing subject is one of the concerns of Herman Rapaport's essay, "Staging: Mont Blanc." Rapaport is interested in the phantasm and its framing, or staging, of a visual scene by desire. In emphasizing the vicissitudes of seeing, Rapaport discusses our necessary misperception (Jacques Lacan's *méconnaissance*) of what cannot be directly looked at—what we fear to let ourselves see and know, and yet cannot not know. In Rapaport's view, Mont Blanc is for Shelley an image of "some kind of horrible power that cannot be precisely specified, looked at, determined." It can only be looked at "ob-scene," with its root meaning of "away from." The seeing subject is displaced from the scene. The spectator is displaced by the theatrical spectacle he stages, yet without the staging he would not be able to see at all.

What are the limits of representation? How can we represent the unrepresentable? In philosophy, Plato can represent the "pure forms" only by a theatrical staging in the famous allegory of the cave in Book VII of *The Republic*. In literature, death is the unrepresentable. Here Rapaport joins Ulmer in looking to Derrida's "Living On: Border Lines" as his model text. In this text Derrida reads Shelley by way of the French novelist Maurice Blanchot. Rapaport's reading of Shelley proceeds in part by way of Derrida's reading of Blanchot's idea, in his novel *L'Arrêt de Mort*, of an *arrêt de mort* (undecidably death sentence *or* reprieve, stay of execution) that is also an *arête de mort* (ridge or arris, hence a kind of stage, of death).

Rapaport then turns to psychoanalysis, in which the unrepresentable is the mother's body. "Staging: Mont Blanc" elaborates an intertextual chain of psychoanalytic references that is extraordinarily rich. Rapaport begins with Freud's interpretation of the Wolf Man's fantasies of the primal scene and elucidates the Wolf Man's "framing" or staging of desire by way of

Catherine Clément's essay (extending a hint of Jacques Lacan) on the theatricality of the phantasm. In support of the Blanchot/Derrida/Shelley material on the undecidability of life and death, Rapaport adduces very pertinent clinical findings on obsessional neurosis from the work of the Lacanian psychoanalyst Serge Leclaire.

In this preface there is space only to gloss two of Rapaport's psycho-analytic citations: Bertram Lewin on the "dream screen" and Jean Laplanche on the genesis of sexuality in the "anaclitic." First, Lewin. In Plato's allegory of the cave, the cave image acts as a stage, a site for representation, of which the perceiver is not conscious. Just so, according to Lewin, the dream screen is a dream without discrete visual content which consists of a blank background that the dreamer does not see as such. The blank screen is the "stage" of the dream. Every dream is pro-jected onto it, but the screen is usually unperceived by the dreamer be-cause it symbolically represents the unrepresentable, the maternal breast as object of desire. This is the breast as hallucinated by the infant during the sleep which follows feeding. Here we see the passive-regressive aspect of the desire, which Rapaport highlights in connection with the "still, snowy, and serene" Alpine landscape. As Laplanche and Pontalis note in *The Language of Psychoanalysis*, "In certain dreams (blank dreams) the screen appears by itself, thus achieving a regression to primary narcissism." So, in the regressive satisfaction of a desire *via* the dream screen, the unrepre-sentable is represented—by a blank. Shelley's mode of representation is similar. He characteristically makes one image the stage or screen upon which another is viewed. The effect, in Shelley's poetic landscapes such as "Mont Blanc," is not of direct perception but of a phantasm. Mont Blanc becomes the veil of life and death, the veiled woman, alluring and terrify-ing, a phantom object rather than a postcard scene.

Mont Blanc is also a prop. It is an object on which the drives lean, "anaclitically." *The Language of Psychoanalysis* notes in its definition of "anaclisis" that "the sexual instincts . . . become autonomous only second-arily, depend(ing) at first on those vital (biological) functions which fur-nish them with an organic source, an orientation and an object." The infant sucking at the breast, as Freud says, is "the prototype of every relation of love." But what is the relation between the breast as object of (biological) hunger and the breast as object of (sexual) desire?

Sexuality is propped on the vital function of feeding, but it achieves autonomy only with the *loss* of the maternal breast, "*in the movement which dissociates it* from the vital function." This view of sexuality stresses the Lacanian theme of *lack*. But no less important is the role of *displacement*, in the relation between the real breast and the phantom breast. The original object is lost and adult sexuality becomes a dream of recovering it. The finding of the sexual object, as Freud says, "is in fact a re-finding of it." But, as Laplanche writes in his book *Life and Death in Psychoanalysis:*

> . . . the sexual object is not identical to the object of the function, but is displaced in relation to it; they are in a relation of essential contiguity which leads us to slide almost indifferently from one to the other. . . . "The finding of an object . . . is in fact a re-finding of it." We would elucidate this as follows: the object to be rediscovered is not the lost object, but its substitute by displacement; the lost object is the object of self-preservation, of hunger, and the object one seeks to refind in sexuality is an object displaced in relation to that first object. From this, of course, arises the impossibility of ultimately ever rediscovering the object, since the object which has been lost is not the same as that which is to be rediscovered. . . .

Thus Shelley's Mont Blanc, if it be an image of the maternal breast, as Rapaport suggests, is not an image present to perception, transparent to the eye. It is a phantom scene, a reminder of a lost origin and, as Laplanche says, of "the essential 'duplicity' situated at the very beginning of the sexual quest," and of the Shelleyan quest as well.

Whereas Gregory Ulmer studies Derrida's word-play in its relation to Derrida's "thought," Tom Conley undertakes a similar study as part of an account of Derrida's literary styles. For Conley, Derrida is preeminently a writer, and the emphasis of his essay is on Derrida's recent work, since *Glas* (1974), rather than on his earlier, more exclusively philosophic writings. For English-speaking readers who know Derrida's work mainly from *Of Grammatology* and *Writing and Difference*, both of which first appeared in France in 1967, Conley's essay will suggest how far Derrida has moved toward becoming a man of letters—both an alphabetic and a literary man.

Ulmer uses the word *Moira* as a prop for studying Derrida's idea that language is destiny. Conley works similar changes on *écart* and its mirror-image, *trace. Écart* (interval, deviation) opens onto *carte* (map), *charte* (deed, title), *écartelement* (quartering), and a number of related terms. This dissemination is what Derrida calls "a systematic and playful exploration of the interval." Exploring *écart*, a word that derives from the Latin *quartus*, enables Conley to show how Derrida dislodges the quadrilateral frame (*cadre*, another derivative of *écart*) of Western perspectival painting and, more generally, the dominant system of representation in Western art and thought.

There is no need here to gloss *écart* because Conley's entire essay explores this space of difference. *Trace*, however, could use some explanation. The word appears in Derrida's important essay of 1968, "Différance," in which he quotes this passage from Heidegger's *Holzwege:* "The matinal trace *(die frühe Spur)* of difference effaces itself from the moment that presence appears as being-present." The reader should keep an eye on that German *Spur* (trace) because it is the spore that ten years later produces *Spurs*, now the English word, used to translate *Éperons*, Derrida's study of Nietzsche's literary styles. In the latter text, Derrida writes that "the English *spur*, the *éperon*, is the 'same word' as the German *Spur:* trace, wake, indication, mark" (p. 41).

Éperons are the spikes attached to the rider's heels and used to urge the horse forward. Since this is the title of a text as much about the question of woman as about the question of style (from which, in Derrida's view, it is inseparable), "spurs" would seem to raise questions about the sexual politics of *Éperons*. Later in this volume, Gayatri Spivak takes up these questions in her essay "Displacement and the Discourse of Woman." What is more pertinent here is the cluster that links writing and style to the various meanings of "spur": trace, track, wake, furrow, mark.

Derrida takes up *trace* in a number of contexts, besides the "Différance" essay already cited: in his deconstruction of Saussurian linguistics in *Of Grammatology;* in his discussion of Freud's use of the mystic writing pad as a model of psychic functioning in *Writing and Difference;* and in *Positions.* The word is one of Derrida's undecidable terms ("not more ideal than real, not more intelligible than sensible, not more a transparent signification than an opaque energy"), and thus eludes any precise definition. The main thing to bear in mind is its relation to *différance* and Derrida's critique of the metaphysics of presence.

For convenience, it may be simpler to cite Jonathan Culler's exposition of *trace* in his recent book, *On Deconstruction.* His example concerns the flight of an arrow; it is one of Zeno's paradoxes.

> At any given moment (the arrow) is in a particular spot; it is always in a particular spot and never in motion. We want to insist, quite justifiably that the arrow *is* in motion at every instant from the beginning to the end of its flight, yet its motion is never present at any moment of presence. The presence of motion is conceivable, it turns out, only insofar as every instant is already marked with the traces of the past and future. Motion can be present, that is to say, only if the present instant is not something given but a product of the relations between past and future.

Another of Culler's examples involves signification. The sound sequence *bat* is able to function as a signifier because it stands in a differential relation with *pat, mat, bad, bet*, etc. The point is that

> The noise that is "present" when one says *bat* is inhabited by the traces of

forms one is not uttering, and it can function as a signifier only insofar as it
consists of such traces.

For Tom Conley, *trace* is inseparable from *écart*, which signifies both
gap and deviation from some limit. One of these limits is the history of
Western writing, its traces from Plato to Freud and beyond. Derrida's
own styles are formed in part on the basis of their differential relations to
the styles of those about whom he writes. So are his styles influenced also
by his *seeing* language and engaging in graphic rhythms that always have a
differential relation with sound. If Conley's own style sometimes seems
hermetic, it is partly because he is trying to reproduce in English Derrida's sensible relations with the French language. Conley is trying to get
the reader to see Derrida's words and screen them much in the way that
Freud handled words in *The Psychopathology of Everyday Life* and in *The
Interpretation of Dreams*.

Altogether, "A Trace of Style" is a Derridean saturnalia of language
play, emphasizing both the visual and the musical aspects of Derrida's
way with words, his achievement of "polyphony through the inscription
of many traditions and voices in single words and letters"—his sensible
language.

MARK KRUPNICK

OP WRITING
Derrida's Solicitation of Theoria

***

Gregory Ulmer

"BUT METAPHOR IS NEVER INNOCENT. It orients research and fixes results."[1] This statement from one of Derrida's early essays summarizes one of the organizing assumptions of his work. Perhaps his best-known application of this Nietzschean suspicion is his study of "writing" as a metaphor: "There remains to be written a history of this metaphor, a metaphor that systematically contrasts divine or natural writing and the human and laborious, finite and artificial inscription."[2] His work searches for alternatives: "One must simultaneously, by means of rigorous conceptual analyses, philosophically *intractable, and* by the inscription of marks which no longer belong to philosophic space, not even to the neighborhood of its other, displace the framing, by philosophy, of its own types. Write in another way."[3] Derrida's examination of certain other founding metaphors—philosophemes—of the Western tradition is as important to his search for alternatives as it is to his deconstruction of logocentrism. He examines *theoria* and *eidos*, theory and idea, in which the sense of sight is sublated to mean thinking itself. What interests Derrida here, what especially calls his attention to this sublation, is the homonym in *sens* (sense): "This divergence between sense (signified) and the senses (sensible signifier) is declared through the same root *(sensus, Sinn)*. One might, like Hegel, admire the generosity of this stock and interpret its hidden sublation speculatively and dialectically; but before using a dialectical concept of metaphor, it is necessary to investigate the double twist which opened up metaphor and dialectic by allowing the term *sense* to be applied to that which should be foreign to the senses."[4]

In this instance, as with other aspects of Derrida's attempt to overcome metaphysics, Heidegger indicates what is at stake. In the statement "Modern science is the theory of the real," what, Heidegger asks, does the word "theory" mean? It stems from the Greek *theorein*, he explains, which grew out of the coalescing of *thea* and *horao*: "*Thea* (cf. theatre) is the outward look, the aspect, in which something shows itself. Plato names this aspect in which what presences shows what it is, *eidos*. To have seen this aspect,

29

eidenai, is to know." And the second root, *horao*, means, "to look at something attentively, to look it over, to view it closely." When translated into
Latin and German, *theoria* became *contemplatio*, which emphasizes, besides
passivity, the sense of "to partition something off into a separate sector
and enclose it therein," from the Latin *templum*, "a sector carved out in the
heavens and on the earth, . . . the region of the heavens marked out by the
path of the sun"—in short, an entirely different experience from that
conveyed by the Greek, stressing now *temenos:* "*Temnein* means: to cut, to
divide." Derrida shares Heidegger's interest in this philological evidence,
if not his commitment to the restoration of another sense of *theorein—
aletheia*, which is translated as "truth," *Wahrheit, veritas:* "the unconceal-
ment from out of which and in which that which presences, presences."
Or, when both roots are included, theory means "the beholding that
watches over truth."[5] Heidegger, nevertheless, makes clear that
"theory"—the term as well as the activity—is not fixed, is still evolving—
an evolution that might be directed by the manipulation of the metaphor
involved. As Derrida writes in his essay "White Mythology":

> Everything in talk about metaphor which comes through the sign *eidos*,
> with the whole system attached to this word, is articulated on the analogy
> between *our* looking and sensible looking, between the intelligible and the
> visible sun. The truth of the being that is present is fixed by passing
> through a detour of tropes in this system. The presence of *ousia* as *eidos*
> (being set before the metaphorical eye) or as *upokeimenon* (being that under-
> lies visible phenomena or accidents) faces the theoretic organ, which, as
> Hegel's *Philosophy of Fine Art* reminds us, has the power not to consume
> what it perceives, and to let be the object of desire. Philosophy, as a theory
> of metaphor, will first have been a metaphor of theory. (*WM*, pp. 55–56)

We are familiar now with the condition of language—its dependency on
such metaphors—which requires a "deconstructive" operation in order to
write "otherwise." As Derrida affirms: "Borges is correct: 'Perhaps universal history is but the history of several metaphors.' Light is only one
example of these 'several' fundamental 'metaphors,' but what an example!
Who will ever dominate it, who will ever pronounce its meaning without
first being pronounced by it? What language will ever escape it?" (*WD*,
p. 92). Associated with the metaphor of light are all the notions of inside-
outside, spatiality and form: "for one would never come across a language
without the rupture of space. . . . For the meanings which radiate from
Inside-Outside, from Light-Night, etc., do not only inhabit the pro-
scribed words; they are embedded, in person or vicariously, at the very
heart of conceptuality itself. . . . No philosophical language will ever be
able to reduce the naturality of a spatial praxis in language" (*WD*, p. 113).

Derrida's response to this situation is at once bold and radical, yet practical and obvious. Finding that no philosophical or scholarly use of language can escape the "nonspeculative" ancestry nor the system of language in use in "civil" or ordinary society (which inevitably introduces "a certain equivocality into speculation"), Derrida concludes: "Since this equivocality is original and irreducible, perhaps philosophy must adopt it, think it and be thought in it, must accommodate duplicity and difference within speculation, within the very purity of philosophical meaning" (*WD*, p. 113). Derrida had already chosen this direction in his first book. There, in his Introduction to Edmund Husserl's "Origin of Geometry," Derrida contrasted Husserl's endeavor "to reduce or impoverish empirical language methodically to the point where its univocal and translatable elements are actually transparent," with James Joyce's literary style, opting for the latter: "To repeat and take responsibility for all equivocation itself, utilizing a language that could equalize the greatest possible synchrony with the greatest potential for buried, accumulated, and interwoven intentions within each linguistic atom, each vocable, each word, each simple proposition."[6]

It is important to realize that much of what Derrida does in subsequent texts is precisely the fulfillment of this program—to build a rigorous, systematic discourse based on the equivocity of language. He has thus produced a metaphorics based on the homonym, homophone, and pun, the structural features of language excluded from metaphor, or the concept of metaphor in classical philosophy. In what follows I will explore how this equivocity functions in the deconstruction of the metaphor of theory.

ARTICULATION

Derrida writes in "White Mythology" that the metaphorization at work in the traditional philosophemes consists of two processes—idealization and appropriation. Against the *Aufhebung* involved in these processes, lifting the sensible into the intelligible (essential to all philosophies of consciousness), Derrida poses a different process—articulation, the sheer syntactic iteration or spacing in language, the hinge of *différance* that joins and separates at once. As a procedure of metaphorization, articulation is "non-Aristotelian" in that Aristotle allowed no place in "metaphor" for *différance*, the joint of spacing.

> For human language is not uniformly human in all its parts to the same degree. It is still the criterion of the noun which is decisive: its literal elements—vocal sounds without meaning—include more than letters alone. The syllable too belongs to *lexis*, but of course has no sense in itself. Above all there are whole "words" which, though they have an indispensable role

in the organization of discourse, remain nonetheless quite devoid of sense, in the eyes of Aristotle. Conjunction (*sundesmos*) is a *phonè asemos*. The same goes for the article, and in general for every joint (*arthron*), everything which operates *between* significant members, between nouns, substantives, or verbs. A joint has no sense because it does not refer to an independent unit, a substance or a being, by means of a categorematic unit. It is for this reason that it is excluded from the field of metaphor as an onomastic field. From this point on, the anagrammatic, using parts of nouns, nouns cut into pieces, is outside the field of metaphor in general, as too is the syntactic play of "joints." (*WM*, pp. 40–41)

Against Aristotle's influential doctrine that "in non-sense, language is not yet born," Derrida builds an alternative system based precisely on what Aristotle (and the tradition after him) excluded from metaphor.

The extent of Derrida's non-Aristotelian inspiration may be seen in Aristotle's condemnation of homonymy as the figure that doubled and thus threatened philosophy. One of the first "places" to check for the obscurity that characterizes "bad" metaphors, according to Aristotle, is to determine whether the term used is the homonym of any other term (*WM*, pp. 53, 74). Derrida, with his interest in discerning and then transgressing the limits of philosophical discourse, takes his cue from Aristotle and builds an entire philosophical program on the basis of the homonym (the homophone, pun, and related devices). In this respect Derrida resembles the nineteenth-century mathematicians who, challenged by the axiomatic absoluteness of Euclid's principles, were able to prove that it was possible to devise a geometry that Euclid's system held to be impossible. Considered at first to be playful monstrosities or abstract exercises, these non-Euclidean geometries eventually provided the mathematics of relativity (just as Derrida's non-Aristotelian grammatology is providing the writing of relativity).

Against the traditional process of sublation dependent on a systematic exclusion of properties for the building of concepts—gathering properties into sets of terms based on synonymy or resemblance of meaning (identity, identification) at the level of the signified, Derrida proposes a homophonic procedure which, working at the level of the signifier, blows a hole (dehiscence) in the cartouche-like boundaries of conceptual categories, allowing terms to circulate and "interbreed" on the basis of the repetition of certain letters (a festival of equivocity). The dehiscence of iteration, an economy that redistributes the property or attributes of names, is exemplified in its generalized mode in "Dissemination," an essay which, as Derrida has explained in an interview, is a systematic and playful exploration of the interval of the gap itself—the term "gap," and the "thing." The term *écart* (gap), that is, permits Derrida in Joycean fashion to generate a series of other terms—*carré, carrure, carte, charte, quatre, trace*—each of which in turn, through its own semantic field, provides him with materials for discussion. He sets to work, within the text of philosophy,

> certain marks. . . , that *by analogy* . . . I have called undecidables, that is, unities of simulacrum, "false" verbal properties (nominal or semantic) that can no longer be included within philosophical (binary) opposition, but which, however, inhabit philosophical opposition, resisting and disorganizing it, *without ever* constituting a third term, without ever leaving room for a solution in the form of speculative dialectics (the *pharmakon* is neither remedy nor poison, neither good nor evil, neither the inside nor the outside, neither speech nor writing; the *supplement* is neither a plus nor a minus, neither an outside nor the complement of an inside, neither accident nor essence, etc.)[7]

These double negations "severely crack the surface of philosophy," contest the logic of noncontradiction while forcing the dehiscence which permits invention by dissemination.

Dissemination offers an entirely new approach to the *idea*, one which alters its status as radically as did Plato's ancient catachresis. Heidegger explains the issue:

> We, late born, are no longer in a position to appreciate the significance of Plato's daring to use the word *eidos* for that which in everything and in each particular thing endures as present. For *eidos*, in the common speech, meant the outward aspect *(Ansicht)* that a visible thing offers to the physical eye. Plato exacts of this word, however, something utterly extraordinary: that it name what precisely is not and never will be perceivable with physical eyes. But even this is by no means the full extent of what is extraordinary here. For *idea* names not only the nonsensuous aspect of what is physically visible. Aspect *(idea)* names and is, also, that which constitutes the essence in the audible, the tasteable, the tactile, in everything that is in any way accessible. *(The Question of Technology,* p. 20)

Derrida initiated in his very first book his counterattack on the appropriation at work in Plato's *idea:* "But it is also necessary to say of the Idea that it has *no* essence, for it is only the openness of the horizon for the emergence and determination of every essence. As the invisible condition of *evidence*, by preserving the *seen*, it loses any reference to *seeing* indicated in *eidos*, a notion from which it nevertheless results in its mysterious Platonic focus. The Idea can only be *understood* [or heard: *entendre*]" *(Husserl's "Geometry,"* p. 142). Although Derrida did not immediately clarify what was involved in this tampering with the metaphor in *idea*, it is possible to note in the present context that in dissemination the *idea* guides thought not by the forms of sight but by the sounds of puns or the distribution of letters. That this strategy is an explicit alternative to theory as *eidos* (the idea as the sublation of the sensible into the intelligible) may be seen in Derrida's allusion to writing in the "key" of I-D (which he discerned in Mallarmé's texts). "The reader is now invited to count the dots, to follow the fine needlepoint pattern of *i*'s and *ique*'s [*-ic* or *-ical*] which are being sprinkled rapidly across the tissue being pushed by another hand. Perhaps he will be able to discern, according to the rapid, regular movement of the machine, the stitches of Mallarmé's idea, a certain instance of *i*'s and a certain scattering of dice [*d*'s]."[8] As the translator notes here, "The word *idée* [idea] is composed of the two syllables in question here: *i* and *dé* [dé = the letter 'd' and the word 'dice']." The idea put to work hypomnemically, as an alternative to the intuition or direct experience of phenomenology (the *idea*—the letters *i-d* pronounced in French—operating mechanically in the repetition of the signifier, collecting a set of terms containing the letters *i-d*), is not the signified concept but the letters or phonemes of the literal word. That the *i-d* of *idée* (idea) also sound the spelling of Freud's term *Id*, confirming that what is involved in this strategy is an *inventio* of *unconscious thought*, opposed to *living* memory promoted in the philosophies of consciousness, is itself evidence of the "insight" available in language itself. What is done briefly with *i-d* in Mallarmé, of course, can be done with other letters, and on any scale, as Derrida demonstrates in *Glas*, which is written with this method, drawing on the " + L" effect by means of which the entire text is generated.

MOIRÉ-MOIRAE

Derrida gets his *ideas* from the systematic exploitation of puns, utilized as an *inventio* to suggest nondialectical points of entry for the deconstruction of the philosophemes. His best-known version of this strategy involves the deflation of proper names into common nouns (antonomasia), as in *Glas*, in which Genet's texts are discussed in terms of flowers (the flowers of rhetoric), beginning with *genêt* (a broomflower). Blanchot, Hegel, Kant, and Ponge have all received similar treatment, described as research into the *signature* effect (itself a punning response to semiology—a strategy of signing in place of a theory of signs). Discussing this methodology in an essay on Ponge, Derrida exposes his mood: "It is necessary to scandalize resolutely the analphabet scientisms . . . before what one can do with a dictionary. . . . One must scandalize them, make them cry even louder, because that gives pleasure, and why deprive oneself of it, in risking a final etymological simulacrum."[9]

The technique works as well for concepts, both for subverting old ones and for building new (pseudo) concepts. Part of my discussion of the critique of *theory* as metaphor is to discern the homophone which (in retrospect, as an aftereffect at least) could be said to be the organizing articulation of Derrida's approach to this project. This search may result in the formulation of an aspect of deconstructive writing which as yet has found few, if any, imitators. The *idea (i-d)* accounting for the specific terms used to deconstruct *theoria* has its source in the "constellation" O-I-R, originally discerned in Mallarmé. (It is worth noting that *oir* is the Spanish equivalent of *entendre*, meaning to hear and to understand, a propos both of Derrida's Joycean macaronics, and his suggestion that the *idea* itself could not be *seen* but only heard.)

> A hymen between chance and rule. That which presents itself as contingent and haphazard in the *present* of language . . . finds itself struck out anew, retempered with the seal of necessity in the uniqueness of a textual configuration. For example, consider the duels among the *moire* (watered silk) and the *mémoire* (memory), the *grimoire* (cryptic spell book) and the *armoire* (wardrobe). (*Dis*, p. 277)

What especially interests Derrida here is precisely the *articulation:* "Rhyme—which is the general law of textual effects—is the folding-together of an identity and a difference. The raw material for this operation is no longer merely the sound of the end of a word: all 'substances' (phonic and graphic) and all 'forms' can be linked together at any distance and under any rule in order to produce new versions of 'that which in discourse does not speak" (*Dis*, p. 277). Derrida is interested in the way in which the arbitrarily rhyming terms have some motivated relationship. To perceive the motivation of the series of O-I-R words for the deconstruction of *theoria* requires that I add one more term to the sequence which Derrida himself neglects, thereby imitating his own addition of *pharmakos* to the series set going in Plato's dialogues: "Certain forces of association unite—at diverse distances, with different strengths and according to disparate paths—the words 'actually present' in a discourse with all the other words in the lexical system" (*Dis*, pp. 129–30). The term is *Moira* (*Destiny* in Greek), a term to which Heidegger devoted an essay, which Derrida has cited. Let us say that the antonomasia, the exchange between proper and common, governing this project, involves *Moirae*— the fates—and *moiré* (not the "watered silk," but the visual illusion known as the moiré effect). *Grimoire* is drawn in with respect to the thirteenth-century fortune-telling book featured in *La Carte postale* (whose wheel of fortune might be associated with Destiny); *memoire* with respect to the artificial memory (hypomnemics) associated with the mechanics of the *inventio*. This *inventio* (an aspect of Derrida's "new rhetoric") functions on the assumption that language itself is "intelligent," hence that homophones "know" something. Derrida's deconstruction of *theoria* reveals what *Moirae-moiré* knows.

Derrida has suggested that the other of *eidos* is *force*, but he rejects any temptation to reduce the problem to a simple dialectical opposition. The situation, rather, is that "force is the other of language without which language would not be what it is" (*WD*, p. 27). This concern for what lies outside of language, the force that is the other of form, is what identifies Derrida's work as "post-structuralist" (indicating that there is something taking place to which the latter term refers):

> In order to respect this strange movement within language, in order not to reduce it in turn, we would have to attempt a return to the metaphor of darkness and light (of self-revelation and self-concealment), the founding metaphor of Western philosophy as metaphysics. The founding metaphor not only because it is a photological one—and in this respect the entire history of our philosophy is a photology, the name given to a history of, or treatise on, light—but because it is a metaphor. (*WD*, p. 27)

Metaphor in general, that is, all analogical displacement of Being, "is the essential weight which anchors discourse in metaphysics." And yet, "this is a fate which it would be foolish to term a regrettable and provisional accident"—as do those who would "cure" language of its "fallen" condition (as in Husserl's striving for univocity). There is a *fate* in language, then, the destiny which resonates in *destinataire*, the recipient of the *envois* in *La Carte postale*. This fate, the determinate of the relationship between force and *eidos*, echoes Heidegger's essay on the idea of Destiny in Parmenides—"Moira."[10] *Moira*, according to Heidegger's analysis of several pre-Socratic fragments, is a force that binds the duality of presencing and that which is present—it is the unfolding of the twofold (in Derrida's terms, the *articulation* of the twofold). As Heidegger remarks, only what is present attains appearance, is receivable, excluding thus from knowledge all that remains concealed, all the rest:

> Destiny altogether conceals both the duality as such and its unfolding. The essence of *aletheia* remains veiled. The visibility it bestows allows the presencing of what is present to arise as outer appearance *(eidos)* and aspect *(idea)*. Consequently the perceptual relation to the presencing of what is present is defined as "seeing." Stamped with this character of *visio*, knowledge and the evidence of knowledge cannot renounce their essential derivation from luminous disclosure. ("Moira," p. 97)

The "fateful yielding" of what is present to ordinary perception by means of "name-words," Heidegger states, occurs "already only insofar as the twofold as such, and therefore its unfolding, remain hidden. But then does self-concealment reign at the heart of disclosure? A bold thought. Heraclitus thought it" ("Moira," p. 100).

Derrida, too, has had this thought—the *idea* of *Moira* becoming the *Moira* of *idea*—not with the same intent as Heidegger, of course (not in the interests of *aletheia*), but in order to locate another way to think *différance*, to name that which does not appear in appearance. In his early analysis of Husserl, Derrida found that structure to be what limits Husserl's attempt to found a phenomenology: "It is not by chance that there is no phenomenology of the Idea. The latter cannot be given in person, nor determined in an evidence, for it is only the possibility of evidence and the openness of "seeing" itself. . . . If there is nothing to say about the Idea *itself*, it is because the Idea is that starting from which something in general can be said" (*Husserl's "Geometry,"* pp. 138–39). In his own system developed subsequently, Derrida calls this condition "writing." Derrida is redefining the idea, working on its root metaphor of sight and light, analyzing it no longer in terms of its effect (the light bulb that lights up when we have an idea in cartoons and advertisements) but in terms of its physics, energy waves (the vibrations mediated by air, the level at which light and sound are equivalents, identified in relation to the body in terms of the "objective senses" of sight and hearing). What electricity is to light, *Moira* is to language. To think grammatologically is not to have an *idea*, but (so to speak) to have a "moira."

What is taking place in Derrida's texts comes into view here in the notion that "light is menaced from within by that which also metaphysically menaces every structuralism: the possibility of concealing meaning through the very act of uncovering it. *To comprehend* the structure of a becoming, the form of a force, is to lose meaning by finding it. The meaning of becoming and of force, by virtue of their pure, intrinsic characteristics, is the repose of the beginning and the end" (*WD*, p. 26). Heidegger also notes this relation between activity and rest: "What is the significance of the fact that destiny releases the presencing of what is present into the duality, and so binds it to wholeness and rest?" ("Moira," p. 98).

Since "force" cannot itself appear, Derrida's strategy is to treat the "binding destiny" that limits what it makes possible by asking, "How can force or weakness be understood in terms of light and dark?" (*WD*, p. 27). The purpose of this question (and here is the principal statement of method) is to "solicit" the founding metaphor of light and sight:

> Structure then can be *methodically* threatened in order to be comprehended more clearly and to reveal not only its supports but also that secret place in

which it is neither construction nor ruin but lability. This operation is called (from the Latin) *soliciting*. In other words, *shaking* in a way related to the *whole* (from *sollus*, in archaic Latin "the whole," and from *citare*, "to put in motion." (*WD*, p. 6)

We are prepared now to appreciate the homonymic, homophonic event (the dehiscence or iteration or articulation that operates in his new metaphorics) which organizes the deconstruction of theory, associating *Moira* (as one of the three elements—destiny, force, and form—included in Derrida's program, as opposed to literary criticism which attends only to form "in every age") with moiré. No concept is available with which to address *theoria* as a philosopheme: "All the concepts by which *eidos* or *morphé* could be translated and determined refer back to the theme of *presence in general*. Form is presence itself. . . . Although the privilege of *theoria* is not, in phenomenology, as simple as has sometimes been claimed, although the classical theories are profoundly re-examined therein, the metaphysical domination of the concept of form cannot fail to effectuate a certain subjection to the look."[11] Hence, Derrida will not use a concept for his deconstruction, but a movement. The liability of language and structure may be solicited in the same way that engineers, using computer analyses of moiré patterns, examine buildings (or any structure) for defects. Just so, the cracks and flaws in the surface of philosophy may thus be located. Of course, an interrogation of this vibration or trembling (by means of which the *Moira* articulating force and form may be re-marked) is the same analogy of thought to the wave motion of light and sound used, Derrida notes, by Hegel in his *Aesthetics* (*WD*, p. 100), Hegel being both the last philosopher of the Book and the first philosopher of Writing. In any case, Derrida is not certain, in his early essays, that any movement other than that of light and sound is possible: "If light is the element of violence, one must combat light with a certain other light, in order to avoid the worst violence, the violence of the night which precedes or represses discourse" (*WD*, p. 117). Borges, he adds, is right again when he says that "'universal history is but the history of the diverse *intonations* of several metaphors'" (*WD*, p. 92), which is why Derrida is still writing about "tone."[12]

CONSTRUCTIVISM

The homophonic principle of articulation suggests, in the manner of an oracle, that *Moira* (the Destiny in language) has something to do with the moiré pattern. What I want to show is that the syntactic movement which Derrida opposes to form is in principle a linguistic version (a "transduc-tion") of the moiré phenomenon being researched by the Constructivist artists in Paris about the same time as Derrida was writing his first book—on the philosophy of Geometry. That Derrida names his soliciting procedure "de-construction" may, in this context, be taken as both an acknowledgment of his affinity with Constructivism, and his caveat that his use of the moiré figure (the intonation that he gives to this metaphor) is not itself another concept of form. Moreover, the story I wish to relate here has nothing to do with influence, but is itself an analogy.

To understand how Derrida carries out his solicitation of the eye's contribution (as founding metaphor) to thought, it is helpful to consider the analogy between grammar and geometry, both of which superimpose figures, one on the lexical and the other on the pictorial world.[13] Geometry, in other words, is the helping science for articulation (just as psychoanalysis is the helping science for other aspects of Derrida's program—even though both are deconstructed in the process). The analogy between grammar and geometry marks the abstracting power of both systems—especially their respective capacities for defining *relations* between objects or words without regard for their specific embodiments or meanings. Geometry and grammar, that is, function at the level of the concept, which in modern thinking is understood *as a set of relations* rather than as a common substance inhering in a group of phenomena.[14] It is not surprising, considering that one of the principal goals of grammatology is to break with the logocentric model of representation, in which writing has been conceived as a representation of speech, that Derrida should look to the non-objective movements in the arts for models of how to proceed.

The Constructivist movement, with its origins in Cézanne (one of whose letters is the point of departure for *La Vérité en peinture*, Derrida's book on the visual arts), and including especially the "Group for Research in the Visual Arts" (GRAV—founded in France in 1960 by Julio le Parc), offers a geometric analogy for grammatological experiments with the non-objective. The artists of GRAV, inspired by the work of Victor Vasarely (active in Paris during the 1950s and later), developed the style which has been dubbed "Op Art"—the creation of optical effects through the manipulation of geometric forms, color dissonance, and kinetic elements, all exploiting the extreme limits of the psychology of optical effect or visual illusion, thus continuing the Constructivist interest (manifested as early as Cézanne and the Cubists) in the interdependence of conception and perception. The experimental production of optical illusion directly in abstract forms (rather than indirectly, as in the mimetic tradition, in forms subordinated to representational demands), is relevant to an understanding of Derrida's attempt to identify the illusory effects of grammar in a similarly pure way. Researching Nietzsche's insight that grammar is the last refuge of metaphysics ("I am afraid we are not rid of God because we still have faith in grammar," Nietzsche said in *Twilight of the Idols*), Derrida inventories some of the irreducible deceptions that grammar generates in our conceptual system (metaphor, for example), demonstrating these effects in exercises that are the grammatical equivalent of the geometrical experiments of the Constructivists working at the limits of optical perception. Derrida's experiment deserves the label "Op Writing" because he is soliciting precisely the concept of theory as a sublation of the metaphors of sight and light.

Op Art provides a guide for an examination of the "trembling" or "shaking" effect which Derrida wishes to achieve in his solicitation of the "idea" as "form" (what does it mean to make a conceptual name tremble?). One of Vasarely's chief techniques, for example, was the development of a "surface kinetics" which set a two-dimensional surface into an apparently three-dimensional pulsation: "In black-and-white patterns—parallel bands, concentric circles or squares, chessboard patterns and the like— pulsating effects are the commonest. A disturbing element, for instance a diagonal or curved line crossing a pattern of stripes, may be added to produce the 'moiré effect.' "[15] Two structures that are superimposed but separate, two different line systems or a line system and a color surface, will also generate the moiré effect. Similar kinetic or disturbing effects are produced by irradiation, the spread of a color beyond its actual surface area. The "Mach strip" or edge contrast on either side of a line dividing two adjacent color areas produces a flutter or vibration along the line depending on the relationship of the two colors. These optical effects are only a few of those taken over by the artists from color theory and cognitive psychology or developed in their own research. But while noting the general analogy between their interests and those of Derrida having to do with framing, grids, networks, movement, and double bands, I want for now to consider specifically how Derrida achieves the moiré effect in language (working with concepts rather than percepts). The moiré effect alone serves not only as a pedagogical model for "solicitation," but constitutes—by virtue of its peculiar feature of being a static form that produces the effect of motion—an emblem of *Moira*, whose nature is to be at once the motion of Becoming and the rest of Being.

Derrida has stated, "no matter what I say, I seek above all to produce effects."[16] The specific effect he seeks, solicitation, turns out to be the textual equivalent of the moiré effect, whose pattern is woven into language on the loom of fate. As already noted in terms of his interest in the ideographic or non-phonetic features of writing (discussed in *Of Grammatology*), Derrida wants to restore to writing the balance between design and symbol (between arbitrariness and motivation, chance and necessity) it had in hieroglyphics. His pursuit of the moiré effect, as an attempt to write the structurality of structure, contributes to this project by experimenting with ornamentation (abstract designs) as formulas or generative devices for text production. In the history of logocentrism, that is, ornament is to geometry what rhetoric is to grammar. As a part of his reversal and displacement of dialectical ideology, Derrida demonstrates the power of thought residing in "decorative" devices.

The moiré effect manifests itself in the special functioning—the equivocality—of Derrida's terminology, illustrated by the term *différance* itself. The verb "to differ" *(différer)* differs from itself in that it conveys two meanings: "On the one hand it indicates difference as distinction, inequality, or discernibility; on the other, it expresses the interposition of delay, the interval of a *spacing* and temporalizing." He concludes that "there must be a common, although entirely different *(différante)* root within the sphere that relates the two movements of differing to one another. We provisionally gave the name *différance* to this sameness which is not *identical*" *(Speech and Phenomena*, p. 129).

Clarity and distinctness are part of philosophy's founding opposition between the sensible and the intelligible (themselves qualities of "literality" suggested by the clarity and distinctness of the alphabetic letter).* As opposed to that seeming clarity and distinctness, *différance* marks a *movement between two letters*—"e" and "a," a "marginal" difference—and between two "differences," a movement that articulates a strange space, "*between* speech and writing and beyond the tranquil familiarity that binds us to one and to the other, reassuring us sometimes in the illusion that they are two separate things" *(Speech and Phenomena*, pp. 133–34). Such, too, is the marginal movement binding *Moira* and moiré.

The strategy of paleonymy (the science of old names) extends this "beat" or rhythm set in motion by the proximity of two meanings, two spellings that are the same and different—an offset overlap of semantic fields within the confines of one term, *like the two overlapping but not quite matching grids that generate the flicker of the moiré effect.* Deconstruction, as a "double science," is structured by the "double mark," by means of which a term retains its old name while displacing the term (only *slightly* or "marginally" at first) toward a new family of terms: "The rule according to which every concept necessarily receives two similar marks—a repetition without identity—one mark inside and the other outside the deconstructed system, should give rise to a double reading and a double writing. And, as will appear in due course: a *double science*" *(Dis,* p. 4). He wants us

---

However, "often in discourse the first or most obvious meaning is quite complex or even more or less purposefully occluded. The fixity of space, however, and the possibility of segmentation suggested by 'literal,' continues to foster the contrary impression, namely that literal meanings, meanings according to the letter, are all fixed and neatly segmental too."[17]

to "see double," hence to have blurred vision. Derrida elaborates this *marginal* structuration—understood as the pattern of superimposed but *slightly* offset grids—as a mime of the history of philosophy itself, which, in spite of its insistence on clarity, is built on blurred repetitions, as in the exemplary case of the distinction between Platonism and sophistics: "And this discrimination itself becomes so subtle that eventually it separates nothing, in the final analysis, but the same from itself, from its perfect, almost indistinguishable double. This is a movement that produces itself entirely within the structure of ambiguity and reversibility of the *pharmakon*" (*Dis*, p. 112). This movement is the grammatological moiré effect.

The moiré effect in Op Writing may be achieved by *marginal* spelling differences, either in the collection of terms on the basis of the repetition of a constellation of letters (as in the O-I-R series), or in the form of anagrams. Describing, or providing instructions for, the reading of *Glas*, Derrida notes, "Each cited word gives an index card a grid *(grille)* which enables you to survey the text. It is accompanied by a diagram which you ought to be able to verify at each occurrence."[18] The terms, that is, are written in the "key" of gl, which is the generative grid for the text. The term Derrida uses to name the virtual movement in *Glas* is *la navette*—the shuttle, referring to the "to and fro" motion which bears this name in weaving, sewing, and transportation. "It is the term I sought earlier in order to describe, when a gondola has crossed the galley, the grammatical to and fro between *langue* [language, tongue] and *lagune* [lagoon] *(lacuna)*" (*Glas*, p. 232). In short, the grids involved are the two spellings, the anagram, with only one letter out of order between them. The shuttle motion between these two words is the binding necessity of their chance occupation of the same letters. The motion is set up within the term "shuttle" *(navette)* itself, joining its meanings or semantic domains, which include in French besides the meanings already mentioned, a liturgical sense (it is a small vessel for incense) and a botanical sense (it is the name of both a plant in the family of crucifers and its seed). "To and fro woven in a warp *(chaîne)*. The woof *(trame*—also *plot)* is in the shuttle. You can see all that I could do with that. Elaboration, isn't it a weaver's movement?" (*Glas*, p. 233). But Derrida states that he distrusts this textile metaphor because it retains a "virtue" of the natural, the originary, of propriety. He decides instead to think of the motion of *Glas* as the interlacing stitching of *sewing*. Both motions evoke the hand-eye relation of writing, as well as the to and fro motion of articulation.

Derrida is also concerned with the way the shuttle motion (the soliciting vibration) is manifested in other systems of thought, especially in psychoanalysis. Freud's famous anecdote of the game his grandson played with a bobbin on a string (the bobbin itself being part of the apparatus of weaving and sewing, symbolizing in this moment of language acquisition the mother, whose loss is repaired with the *fort-da* stitch), serves Derrida as the pretext guiding his reading of *Beyond the Pleasure Principle*. In fact, the two dominant structuring devices in *La Carte postale* repeat emblematically the *Moira*-moiré homophone. The first half of the book, the "Envois" section, is written around the postcard with its representation of the frontispiece of a medieval fortune-telling book *(Moira)*; the second part of the book contains the monograph-length "Speculer—sur 'Freud,'" in which Derrida reads the *oscillating motion* of the *fort-da* scene as the structuring principle or design of the entire text (moiré).

In *Glas*, the conceptual equivalent of the back and forth motion of sewing is the undecidability of the fetish, the very topic being treated in the Hegel column at the place in the Genet column at which the "shuttle" is discussed: "Here, he (Freud) comes to recognize the 'fetishist's attitude of splitting' and the oscillation of the subject between two possibilities" *(Glas,* p. 235). The oscillation enters the Genet column later: "he oscillates like the beating of a truth which rings. Like the clapper in the throat, that is to say in the abyss of a bell" *(Glas,* p. 254). *Glas,* having found in the homophonic shuttle a different intonation of one of the philosophemes, sounds the knell of dialectics, and writes otherwise.

GROTESQUERIE

The question of fetishism, "an economy of the undecidable," in the context of soliciting the sensorium—the role of the senses in the root metaphors—concerns membranes, their tension and capacity for vibration or permeability (the tympanum or the hymen). The eye, too, is a membrane, as Derrida notes in citing Mallarmé: "*Gloire du long désir, Idées* ('Glory of the long desire, Ideas') rhymes with *La famille des iridées* ('The iris family'). The iris, the flower absent from all bouquets, is also the goddess of the rainbow and a membrane in the eye" *(Dis,* pp. 284–285). In relation to the body, the association of the light-dark philosopheme with the inside-outside opposition is especially evident. When the fetishist looks at woman unveiled, he sees and denies the absence of the phallus, sees the mother's phallus, model of all simulacra. Working his way out of the paradigm of the idea become woman, Derrida looks for evidence of oscillation in the vicinity of the "vagina" as term. He considers, to begin with, the undecidable nature of the hymen occupying the space *between* *(entre)* the inside and the outside. He discovers the shuttle at work here in the very name of this space—the between, *entre,* since the word *antre* (a cavern or grotto) also names the vagina, and even, etymologically, *entre* itself. The hymen, in Derrida's terminology, is the structurality (the vibration of a membrane) of these two words together. The homophones *entre-antre* enact a repetition of signifiers that is the device constitutive of grammatological "space" in which the root metaphors of *theory* and *idea* can be solicited.

> Without reducing all these to the same, quite the contrary, it is possible to recognize a certain serial law in these points of indefinite pivoting: they mark the spots of what can never be mediated, mastered, sublated, or dialecticized through any *Erinnerung* or *Aufhebung*. . . . Insofar as the text depends upon them, *bends* to them, it thus plays a *double scene* upon a double stage. It operates in two absolutely different places at once, even if these are only separated by a veil, which is both traversed and not traversed, *inter*-sected. Because of this indecision and instability, Plato would have conferred upon the double science arising from these two theaters the name *doxa* rather than *episteme*. (*Dis*, p. 221)

Such is the nature of the *liaison* of the two semantic domains articulated by the jointing of the shuttle. What is only a marginal displacement at the level of the letter, setting up the grammatical equivalent of a blur, reaches catastrophic proportions at the conceptual level, prohibiting the unifying effects, the clarity and distinctness, of dialectics.

That the between is also a *grotto (entre-antre)* is important for understanding the place of grammatology in the history of ornament, since it suggests that Op Writing is a form of grotesquerie. "These grottal effects are usually also glottal effects, traces left by an echo, imprints of one phonic signifier upon another, productions of meaning by reverberations within a double wall. . . . The decisive, undecidable ambiguity of the syntax of 'any more' [*plus de*] (both supplement and lack)" (*Dis*, p. 274). An example on a small scale of Derrida's participation in the genre of the grotesque is "Tympan," the introductory essay in *Marges*. Its topic, relevant to the title of the collection, is the margins and limits of philosophy. The double-column format is used (anticipating *Glas*). The right side is a citation from Michel Leiris, and the left side is Derrida's discussion "touching" on Leiris—marginal writing, writing in the margins (the place for decorations). The citation presents what can best be described as a rhetorical *cartouche*, a version of the *flourish* in ornament—it is an inventory of network, grid, woven or winding patterns found in nature and society, all of which somehow remind Leiris of the name "Perséphone":

> The acanthus leaf which one copies at school when one learns to handle charcoal more or less well, the stems of a convolvulus or other climbing plants, the spiral inscribed on a snail shell, the meanders of an intestine, the curl of childhood hair enshrined in a medallion, the modern style ironwork

of metro entrances, the interlacings of embroidered monograms on sheets and pillow cases, the windings of a path, everything that is festoon, volute, scroll, garland, arabesque. It is a question therefore, essentially, of a name in *spirals.*" (*Marges,* pp. ii–v)

Derrida, no doubt, offers the arabesques as an image of the scribble that is writing. For Leiris, the signature "Perséphone" is inscribed in the framing flourish; the shapes of the spiral sign the name.

As E. H. Gombrich notes in a chapter of *The Sense of Order* entitled "The Edge of Chaos," the margin with its overgrowth of tendrils, in baroque artists like Albrecht Dürer, may spawn *monsters, grotesques* (an extension of the original decorative term) resulting from the playful invention permitted in this "zone of license":

> Much learning and ingenuity has been expended in assigning symbolic meanings to the marginal flourishes, monsters, and other motifs created by Dürer and his medieval predecessors, and there is no reason to doubt that once in a while the text offered a starting point to the artist for his playful invention. But even where we are prepared to strain our credulity, the majority of inventions must still be seen as creations in their own right.[19]

Dürer, the example of this tradition of virtuosity and free invention in marginal decoration, mixed every known tradition in a search for ideas. The resulting enigmatic images are classified as "grotesques" or "drolleries." Dürer's own term was "dreamwork" ("Whoever wants to dream must mix all things together"), creating an effect of "bewildering confusion."

It is useful to consider Op Writing within this tradition of the grotesque. Leiris, in the passage cited in "Tympan," provides an example of how the grotesque, in its original or technical sense (which refers to a kind of ornament, similar to the arabesque, consisting of medallions, sphinxes, foliage), can be extended (in its language version) to the "fantastic" mode of the style, thus combining like Dürer the decorative and the monstrous. The spiraling foliage (decorative grotesque) reminds Leiris of the name Perséphone, but the name itself sounds to him like *perce-oreille*—earwig, a boring insect—which is the appearance of the "monster" (antonomasia, or homophone, as "metamorphosis"). Derrida's own homophonic or punning strategy results in a similar "fantasy etymology," which has much the same distorting effect in a philosophical discourse as had Dürer's drawings of interlacing thistles, cranelike birds, and gargoyles on the margins of the prayer book the Emperor Maximilian had commissioned for his newly founded Order of St. George.

As suggested by *antre*—the grotto, recalling the Italian grottoes in which the ancient decorations were discovered, hence their being dubbed "grotesquerie"—the "betweenness" of grammatological space is a zone of license. Part of the lesson of the grotesque genre for understanding Derrida, keeping in mind Gombrich's stress on the independence of grotesquerie, and all ornamentation for that matter, from what it decorates, is that Derrida's writing deals "only" *marginally* with what it is "about" (with what it surrounds or enframes, like the passe-partout matting), a feature which I will discuss in the context of the figure of the "interlace." Nonetheless, this marginal relation between discourse and "object of study" retains a mimetic quality: the moiré effect of Op Writing, giving rise to grotesque (in the technical sense of the term), homophonic etymologies, constitutes a new theory of mimesis (Derrida is opposed, he says, not to mimesis, but to a determined interpretation of mimesis which he calls "mimetologism" [*Positions*, p. 70]). "Here we are *playing* on the fortuitous resemblance, the purely simulated common parentage of *seme* and *semen*. There is no communication of meaning between them. And yet, by means of this floating, purely exterior collusion, accident produces a kind of *semantic mirage:* the deviance of meaning, its reflection-effect in writing, sets something off" (*Positions*, p. 45). The new mimesis, in short, is based on homophonic "representations."

The explanatory efficacy of the *Moira*-moiré conjunction has persuaded me that Derrida is on to something with his mimetics of the signifier. The metaphorics of non-Aristotelian articulation permits the generation of a discourse between the pulsating moiré effect (emblem of solicitation as "trembling") and *Moira* or Destiny. The hinge joining these two domains may be found within the tympanum itself, whose meanings, as Derrida notes in "Tympan," include of course the vibrating eardrum (sound and light being susceptible to the same effects, the beat of dissonance being the acoustic equivalent of the moiré blur—both effects of proximity), and a type of water wheel—suggesting an image of the "wheel of fortune."

INTERLACING

Relevant to grotesque "tracery," the interlacing pattern—the structure of chiasmus and of "invagination" which constitute the structurality of structure in Derrida's texts—is a major feature of ornament throughout history. The seventeen symmetries producible by the three principles of rhythm (rotation, reflection, translation) can be multiplied, Gombrich explains, to eighty possible arrangements by the simple device of the interlacing of the lines above and below one another, a procedure that introduces the fiction of a mirroring plane. The illusion of depth thus introduced, giving the effect of weaving, plaiting, knotting (cf. the shuttle, the braided fetish), places the interlacing device undecidably between abstract form and representational meaning—between geometry and a perceptual thing, thus providing an analogy (exploited in "Speculer—sur 'Freud' ") for the ambiguous status of *speculation*. A universal device, one of the ones most frequently encountered in designs all over the world, the interlace is also one of the favorite patterns of Op Art (and also of Op Writing, in which it serves as an image of syntax). The moiré effect itself is produced by a kind of interlacing of superimposed grids.

Derrida uses the notion of interlacing to account for the marginal relationship between his discourse and its examples ("theoretical" discourse as a kind of decoration on the borders of another text). He uses the "loophole" of a figure provided by set theory—the modern heir to the notion of the concept as a "having" or "belonging to"[20]—in order to describe the paradoxical escape of the example from conceptualization. The figure is that which Derrida formulates as "the law of the law of genre":

> It is precisely a principle of contamination, a law of impurity, a parasitical economy. In the code of set theories, if I may use it at least figuratively, I would speak of a sort of participation without belonging—a taking part in without being part of, without having membership in a set. The trait that marks membership inevitably divides, the boundary of the set comes to form, by invagination, an internal pocket larger than the whole; and the outcome of this division and of this abounding remains as singular as it is limitless.[21]

The invaginating fold displays graphically the structure of catastrophic metaphors, a term ("catastropic") which, in this context, may be recognized as a further analogy with geometry, alluding to that division of topology known as catastrophe theory. The catastrophic metaphor is one which by invagination scatters rather than collects properties. The "fold" is the simplest of the seven elementary catastrophes—"catastrophe" referring to the *event of discontinuity or instability in a system.*[22] Derrida's most frequent use of the invagination metaphor, describing the hymeneal effect of discontinuity or transformation resulting from his homonyms, is that of turning a glove inside-out (which would transform a right-hand glove into a left-hand glove—hence the discontinuity): "One can always, although it is never indispensable, turn the reference like a glove. Pretending to describe this or that, the veils or sails, for example of saliva, the text veils itself in unveiling itself by itself, describing, with the same exhibitionistic modesty, its own texture" (*Glas*, p. 160).

A glance at a general description of the concerns of catastrophe theory suffices to indicate its relevance to Derrida's interests in boundaries and borderings as they exist in the humanities. "As a part of mathematics, catastrophe theory is a theory about singularities. When applied to scientific problems, therefore, it deals with the properties of discontinuities directly, without reference to any specific underlying mechanism. This makes it especially appropriate for the study of systems whose inner workings are not known."[23] The natural phenomena of discontinuity to which it may be applied include "the breaking of a wave, the division of a cell or the collapse of a bridge, or they may be spatial, like the boundary of an object or the frontier between two kinds of tissue." In short, it is the perfect resource for Derrida's use of the membrane as an analogy, and for writing on margins.

But what most concerns Derrida is the bridge and its potential collapse (the instability of structures tested by moiré patterns)—the bridge of analogy itself, as discussed in Kant's aesthetic of the Sublime:

> The bridge is not *an* analogy. Recourse to analogy, the concept and the effect of analogy are or make *the bridge* itself. . . . The analogy of the abyss and the bridge over the abyss, is an analogy to say that there ought to be an analogy between two absolutely heterogeneous worlds, a third to pass over the abyss, to cicatrize the chasm and to think the gap. In short a *symbol*. The bridge is a symbol.[24]

Kant's model of analogy, part of a powerful tradition still operative today, Derrida notes, is dialectical, based on a certain *continuity* from the known to the unknown and from the concrete to the abstract, allowing innovation to occur by means of proportionality and symmetry. But Derrida is interested in a *discontinuous* model of innovation and change, which he describes in his critique of Condillac, *The Archaeology of the Frivolous*. This model "produces a silent explosion of the whole text and introduces a kind of fissure, rather fission, within each concept as well as each statement," as happens "when the analogy is weak, the 'quantity of connection' not great enough." In these circumstances, analogy *misleads*, becomes frivolous: "A 'stretched' sense always risks being empty, floating, slackened in its relation with the object." But the very structure of the sign—its *disposability* in the absence of the thing—makes frivolity a "congenital breach" in language. The homonym, to be sure, is the most frivolous relation of all because it produces a crossing with the *least* "quantity of connection," being an empty repetition of the signifier: "Frivolity originates from the deviation or gap of the signifier, but also from its folding back on itself in its closed and nonrepresentative identity."[25] Repetition—the folding of the signifier on itself—as such can produce the effect of invagination.

Derrida uses a "genreless" text by Blanchot, *La folie du jour*, to "illustrate" his point. The invagination or folding (a form of the interlace) explored in this story (which Derrida has discussed in several articles) involves a re-citation of the "beginning" of the story at the end in a way that blurs the distinction between discourse and quotation:

> Each story is part of the other, makes the other a part (of itself), each "story" is at once larger and smaller than itself, includes itself without including (or comprehending) itself, identifies itself with itself even as it remains utterly different from its homonym. Of course, at intervals ranging from two to forty paragraphs, this structure of *crisscross double invagination*. . . . never ceases to refold or superpose or *over-employ* itself in the meantime, and the description of this would be interminable. I must content myself for the moment with underscoring the supplementary aspect of this structure: the chiasma of this *double invagination* is always possible, because of what I have called elsewhere the iterability of the mark.[26]

Iterability, the sheer possibility of quotation, or repeating, creates the catastrophic fold in *any text*, giving it the structure of a Klein bottle (in topology, a single surface "with no inside, outside, or edges. It is formed by drawing the smaller end of a tapering tube through one side of the tube and then enlarging the former until it fits the latter"[27]) by opening the inside to the outside. Such a text will not have or hold properties—the figure thus of a text operating against the *concept* as such—any more than a Klein bottle will hold water. The paradoxical topology of the Klein bottle (recalling the pots with holes knocked in the bottom found in tombs, which Kant disqualified from the category of beauty because they manifested a purpose for which they had been rendered useless), represents grammatology's theoretical conversion of the containers described in some histories as constituting the origin of writing (the clay spheres or *bullae*, filled with tokens representing commodities, whose content later came to be depicted on the *outside* by inscriptions, leading finally to the abandonment of the containers altogether, leaving the inscriptions to function independent of their guarantee).[28] The analogy of invagination, in short, is a deconstruction of the notion of language as a "container" for ideas.

Another aspect of the invagination that especially interests Derrida concerns its narrative form: "double invagination, wherever it comes about, has in itself *the structure of a narrative (récit) in deconstruction*. Here the narrative is irreducible. . . . The *narrative of deconstruction in deconstruction*" (*LO*, p. 100). No text can refer to something beyond itself "without becoming double or dual, without making itself be 'represented,' refolded, superposed, *re-marked* within the enclosure, at least in what the structure produces as an effect of interiority" (*LO*, pp. 100–101). Every text, in other words, places *en abyme*—using the "abyss" (of analogy) now in the idiomatic sense given to it in the terminology of the arts—a model of itself. *Mise en abyme*, a term borrowed from heraldry—a figure *en abyme* is located at the heart of the escutcheon, *"but without touching any of the other figures"*—meaning by analogy "any enclave entertaining a relation of similitude with the work which contains it."[29] The example in a critical or theoretical discourse—like the scene of the bobbin in *Beyond the Pleasure Principle*—functions as a kind of figure *en abyme*, analogous in its effect to the mirrors placed strategically in certain paintings, rendering visible what takes place "behind our backs," as in Van Eyck's *Arnolfini and his Bride*, which includes not just a miniaturization of the scene, but shows the painter himself, creating an effect of oscillation between the inside and the outside of the frame.[30]

Thus the enfolding that most interests Derrida is precisely the interlacing chiasmus of the narrator and the narrative with the "content" of his story or discourse (his critique of Lacan's Seminar on "The Purloined Letter" is directed at Lacan's failure to take this interlacing into account). In the essay "The Law of Genre" Derrida provides a simple diagram of the structuration—an interlacing of two curving lines, Op Writing's equivalent of the ornamental weave, described now as "a double chiasmatic invagination of edges" (*LG*, p. 218). The application of the "law of the law of genre" in his own writing—modeling the relation of his discourse to its object of study (examples) on the paradoxical hierarchy of classification in set theory (the "law of participation without membership, of contamination"—*LG*, p. 210)—is apparent in "Living On: Borderlines." The question he poses, faced with the problem of a comparison of Blanchot's *L'arrêt de mort* with Shelley's *The Triumph of Life*, is: "How can one text, assuming its unity, give or present another to be read, without touching it, without saying anything about it, practically without referring to it?" (*LO*, p. 80). His procedure will be, he says, to "endeavor to create an effect of *superimposing*, of super-imprinting one text on the other," a version of "the double band or 'double bind' of double proceedings" used in *Glas*, which breaks with the conventional assumptions of pedagogy: "One procession is superimposed on the other, accompanying it without accompanying it. This operation would never be considered legitimate on the part of a teacher, who must give his references and tell what he's talking about, giving it a recognizable title. You can't give a course on Shelley without ever mentioning him, pretending to deal with Blanchot, and more than a few others" (*LO*, pp. 83–84).

Designed to avoid the dialectical effects of traditional commentary, the technique of "superimposition" of texts recreates at the level of theoretical discourse the solicitation of *theoria* conducted at the level of the philosopheme, retaining the structure of the moiré phenomenon throughout. No bridge of analogy, in other words, no direct confrontation at all, in the manner of dialectics, but only the contaminating blur or virtual flutter of two overlapping forms—such is the procedure for having a moira rather than an idea.

ORNAMENT

A comparison of E. H. Gombrich's study of decorative art, *The Sense of Order*, with Derrida's Op Writing (his use of various devices from the visual arts to investigate the sight metaphors in *theoria* and *idea*), reveals that many of the effects Derrida seeks (only a few of which I have been able to discuss in this article) are those inherent in the history of ornament—decorative or parergonal art—of which Constructivism and other abstract art movements, as Gombrich explains, are the modern heirs. Derrida's research into these decorative devices is a deliberate aspect of his "metaphorology," challenging the logocentric prejudice against rhetoric as "ornament" and showing that ornamentation itself can provide the methodology of a science—grammatology.

Recalling Derrida's exploitation of equivocity, we might also hear in "ornament" the meanings elaborated by Angus Fletcher, which indicate what may be at stake in the application of decorative designs to concept formation. In the history of rhetoric, the term "ornament" was gradually generalized to include all the figures of speech and all tropes.[31] The oldest term for ornamental diction, Fletcher says, is "kosmos," indicating that there is a "world view" at work in the decorative. More specifically, "ornament" and allegory refer to different aspects of the same process, having to do with the relation between part and whole, micro- and macro-"kosmos." The traditional assumption—that the part implies the whole (or vice versa), that part and whole are complementary to one another,[32] is altered by Derrida's invaginated interlace to suggest that part and whole have rather a *supplementary* relationship.[33]

Derrida's interest in the features and history of ornament is evident in his concern for everything marginal, supplementary, everything having to do with borders rather than centers. Gombrich mentions that he thought of calling his book "the unregarded art," since decoration, as parergon or bywork, is not noticed, its effects being assimilated inattentively, with peripheral vision. Against the logocentrism of Western metaphysics which thinks of style as something added on to thought—decoration—and which valorizes the center of structure—the notion of presence which is both inside the structure yet outside, controlling it, out of play—Derrida proposes that our era is beginning to think the structurality of structure, realizing that the center is not a natural or fixed focus but a function, "a sort of non-locus in which an infinite number of sign substitutions come into play."

Replacing the old notion of "center" is the notion of supplementarity, described as a movement of freeplay:

> A field of infinite substitutions in the closure of a finite ensemble. . . .
> instead of being an inexhaustible field, as in the classical hypothesis, instead
> of being too large, there is something missing from it: a center which arrests
> and founds the freeplay of substitutions. . . . One cannot determine the
> center, the sign which supplements it, because this sign adds itself, occurs
> in addition, over and above, comes as a supplement.[34]

The entities used as models for supplementarity (the effect of Enframing) include, for example, the device known as the *passe-partout* (the matting used for the display of prints or engravings, open in the center for the infinitely substitutable image; and also a "master-key"); the *cartouche* (in one sense, the decorative border, whose convolutions may be extended infinitely within the closure) surrounding or framing a blank space ready to receive an inscription (in addition to its meaning in hieroglyphics as the oval shape enclosing the glyphs of the proper name). Both these items or -examples are explored in *La Vérité en peinture*.

Gombrich's study of ornament, then, helps account for Derrida's manner of interrogating framing effects, determining what is included and what excluded (from a concept, for example). Gombrich defines the frame as a continuous break setting off the design from the environment. Nor can there be a center without a frame. Patterning, like ordering of any kind, is the ordering of elements of identity and difference: in ornamentation its two steps are framing and filling. The geometrical tendency in design starts at the outside or frame of the surface and works in to the center (grammatology's approach), while representational (naturalistic) designs tend to begin at the center and work out toward the frame.

In lieu of a conclusion, or by way of opening the topic to further investigation, I would like to point out several other elements discussed in Gombrich's study of ornament that are relevant to Derrida's deconstruction of the optical effects in conceptual discourse (the philosophemes). Reading Gombrich, who notes that pattern is a form of rhythm, made me more aware of what Derrida might mean when he talks about "rhythm" replacing dialectic in a new "theory" of change: "Inseparable from the phenomena of *liaison* . . . the said unities of time could not help but be also metrical and rhythmic values. Beyond opposition, the difference and the rhythm" (*La Carte postale*, p. 435). "Rhythm," that is, is spatial as well as temporal. The laws of repetition (repetition as the surrogate which is not a copy of anything being the principle of decentering—the structurality of

structure, as in Edmond Jabès's repetition of the Book—in Derrida's program) governing pattern formation, Gombrich explains, include *translation* (rhythmic rows extended along an axis), *rotation*, and *reflection*. By
each of these principles, grids or lattice forms may be generated and
extended infinitely.

The mathematical employment of these concepts of rhythm, as James
Ogilvy has demonstrated, serves to map one set of axes onto another
(indeed, Jakobson defines poetry as the projection or mapping of one axis
of language—paradigmatic or syntagmatic—onto the other), making them
useful operations for understanding (or even for bringing about) relation-
ships between the varous dimensions of discourse—for the circulation of
the philosophemes through all the divisions of knowledge:

> The usefulness of the concept of transformation [read translation, rotation,
> reflection] consists in the fact that, unlike the more familiar notion of anal-
> ogy, transformation permits the more radical move toward taking the basic
> parameters themselves—the political, psychological and religious dimen-
> sions—as transforms of one another. Unlike symbolism and analogy, which
> tend to assume a basic or literal foundation on which an analogy is built or a
> symbol drawn, the concept of transformation assumes no fundamental
> dimension.[35]

The purely relational and mathematical operations of ornament, applied
to the conceptual dimension, make irrelevant the notions of proper and
figurative meanings. Ornament, taking into account the *spacing* of grammatology, offers an alternative to analogy.

Op Writing exploits for its effects the tendency to receive concepts in
terms of presuppositions and the encoded habits of expectation, in the
same way that Op Art exploits the fact that the eye "is good in recognizing
continuities and redundancies, but bad in 'locking in' on a particular
feature of repeated elements." One of the assumptions of the solicitation of
theoria by means of Op Writing is that concept formation tends to repeat
the weaknesses of seeing. Thus an art—or a writing—*based on repetition*
will cause problems for the habits of both seeing and thinking: "By strictly
repeating this *circle* in its own historical possibility, we allow the production of some *elliptical* change of site, within the difference involved in
repetition. . . . Neither matter nor form, it is nothing that any
philosopheme, that is, any dialectic, however determinate, can capture"
(*Speech and Phenomena*, p. 128).

As with Vasarely, who took as the basic element of composition (his
plastic "unit") two contrasting color shapes, many Op Art effects play
with the contrasting relation of figure and ground, and the oscillation and
interferences set up in the play between "two bands" (recalling Derrida's

contra-band strategy). Much of this effect is due to an "extrapolator" device in visual perception which goes beyond mere registration to the production of continuous shapes, a performance of habit and anticipation which the artist uses to create the illusion. Gombrich explains that Op Art achieves many of its effects by means of a *systematic overload of the perceptual apparatus*. An example is the "Fraser Spiral," "which is not a spiral at all, but a series of concentric circles superimposed on vortex lines. These lines, it turns out, tend to deflect our searching gaze so that we always lose our place and settle for the most plausible "templet," the continuous spiral."[36] Lightening the load of information by covering half the page or tracing the lines with a pencil reveals the trick.

Derrida similarly plays with our conceptual habits favoring the continuities of common sense, and he overloads our conceptual apprehension with a paradoxical syntax that displaces the normal line of logic, resulting in a conceptual "vertigo" akin to the "Fraser Spiral." The grammatological equivalent of perceptual overloading occurs in terms of the homonymic relation between "translation" of language and "translation" as a procedure of pattern formation in ornament. Derrida, that is, calls attention to this technique (a version of solicitation) while posing an undecidable problem of translation. Discussing in "Borderlines" his treatment of Blanchot in "Living On," Derrida states: "There, too, are economy and formalization, but by semantic accumulation and overloading, until the point when the logic of the undecidable *arrêt de mort* brings and opens polysemia (and its economy) in the direction of dissemination. Why have I chosen to stress the translation-effect here? Effects of transference, of superimposing, of textual superimprinting between the two 'triumphs' or the two '*arrêts*' *and* within each of them" (*LO*, p. 91). The example given here of semantic overloading—*arrêt de mort*, which normally means "death sentence," but which in a literary context could mean "suspension of death"—is just one among many in Derrida's texts. Besides the "*plus de*" (any more)—both supplement and lack—mentioned earlier, some of the other examples mentioned in "Borderlines" include *pas de méthode*—"no method" but also "a methodical step"—and "*point de méthode*—"absolutely no method" but also "a point of method" (*LO*, pp. 96–97).

One reason that Derrida's texts are difficult, or lend themselves to misunderstanding, is that he is working in a non-objective register which is by definition unfamiliar to his readers, since his purpose is to experiment with alternatives to the philosophemes, and even to work not with concepts but with kinetic effects. He does alert us, however, to what he is doing. In "Restitutions," for example, in which the laces in Van Gogh's paintings of shoes become an emblem for the structuring or "stricturing" of the *interlace*, Derrida states: "To think *otherwise* does not imply that one thinks without relation or in a simple relation of transformation altering current or philosophical thought, but still according to another relation of interlacing, which is neither the reproduction nor the transformative production of a given material" (*Vérité en peinture*, p. 403). One of the images he provides in this essay for his organization of the materials (Van Gogh's paintings and essays by Heidegger and by Meyer Schapiro), is another version of the superimposition of the moiré effect: "*Macula* defines the limits, that's what remains to be seen, and, if it's the law for us, we can't meddle with it."[37] "Macula," or "mackle," is defined as "a blur in printing, as from a double impression."

The lability of his own system is emphasized in "Limited Inc," in which the "restlessness" which is often attributed to the figures of Op Art is seen to be a feature of Op Writing as well, as exemplified by the instability of the term *restance* itself. When he originally introduced *restance* in the article "Signature Event Context," Derrida equipped it with several "warning" signals. It is, to begin with, a neologism, translated into English as *remainder*, explicitly avoiding the word *permanence*, with *restance* being retained in brackets. The graphics utilized—the italic print and the bracketed term—serve as a warning light, Derrida explains. In its context, the term is also associated with "non-present" which "adds a spectacular *blinking-effect* to the warning light. . . . Blinking is a rhythm essential to the mark whose functioning I would like to analyze."[38] The "graphics" of *remainder* signal that this term marks a "quasi-concept": "To remain, in this sense, is not to rest on one's laurels or to take it easy" (Limited Inc," p. 190). Such, too, is the lability or restlessness of all of Derrida's terms, which could be said to tremble or flutter as if in a moiré pattern.

At least two approaches (both having to do with topology) to further analysis, and replication, of what I am calling Derrida's moira are available. The first approach involves a "voco-sensory" theory of language, a "glossodynamics," which maintains essentially that word meanings are motivated by the position and movement of the speech organs, especially the tongue—the theory of "articulation" which Derrida plays with in *Glas*. One proponent of this position uses topology metaphorically to describe the Hebraic attitude to writing (which is the real relevance of this point to Derrida's strategy):

> One of the chief characteristics of Semitic word-formation is the preoccupation with consonants as against vowels, which in archaic writing were omitted. . . . The consonants constituted the substance of any word; the vowels formed it, and therefore could change in accordance with the inflection. The word might be pictured as a wire frame or a rubber object which, by manipulation, could take on any form; elongated, flattened, rounded,

etc. The word thus possesses something of the topological properties in mathematics.[39]

Glas, with its " + L" effect, working with the GL "phonex" as a generative grid, is an application of this Semitic principle to grammatology.

Given the importance of prepositions (especially *de* and *sur*, among others) to Derrida's undecidable syntax, producing the effect of semantic overload, the "glossodynamic" description of how prepositions function is also relevant: "both the place adverb and the preposition play the same part, *viz.*, relate one fact to another, which *visually suggests light and rapid movement.*"[40] Moreover, in teaching prepositions to children, for example, "there seems to be no other way of expressing the meaning of prepositions than by *tensions* of various kinds. One cannot define them in terms of anything, because there is nothing else to them but a variety of muscular tension, which when no other words are given with them is quite without ideational content."[41] Derrida's utilization of membrane images—the hymen, tympanum, larnyx, and iris—to remark articulation may be recognized as emblems for the functioning of prepositions in Op Writing.

The other approach to this level of language is "properly," rather than metaphorically, topological. I have in mind the need for further investigation into the applicability of René Thom's catastrophe theory—with its roots in biology rather than physics—to Derrida's deconstructive strategy. Derrida, of course, admits only to using his various mathematical and biological terms metaphorically, but (to repeat his own statement with which I began this paper) "metaphor is never innocent." René Thom, in any case, considers catastrophe theory to be a new theory of analogy, perhaps the first truly "different" approach to analogy we have seen since Aristotle. It might be worthwhile, in the context of a solicitation of *theoria* and *idea*, to experiment with the usefulness of Thom's comparison between language structure and molecular biology, using geometric modeling: "What I propose is nothing less than a vast metaphor between the symbolic of *Structure* . . . and the dynamic of embryology."[42] Perhaps, as Derrida sometimes says, this project will do itself. But such serious proposals should not allow us to forget the basic issue explored here (like forgetting Nietzsche's "umbrella")—the "luck" or *destiny*, between chance and necessity, which renders permeable the membrane dividing life from language and gives us the *Moira Effect*.

NOTES

1. Jacques Derrida, *Writing and Difference*, trans. Alan Bass (1967; Chicago: University of Chicago Press, 1978), p. 17. Hereafter cited in the text as *WD*. After the first full citation of each work by Derrida, all subsequent references to that work appear in the text.

2. Jacques Derrida, *Of Grammatology*, trans. Gayatri Spivak (1967; Baltimore: Johns Hopkins University Press, 1976), p. 15.

3. Jacques Derrida, *Marges de la philosophie* (Paris: Minuit, 1972), p. xx.

4. Jacques Derrida, "White Mythology: Metaphor in the Text of Philosophy," trans. F. C. T. Moore, in *New Literary History* 6 (1974): 27–28. Hereafter cited in the text as *WM*.

5. Martin Heidegger, *The Question Concerning Technology*, trans. William Lovitt (New York: Harper and Row, 1977), pp. 163–65.

6. Jacques Derrida, *Edmund Husserl's "Origin of Geometry": An Introduction*, trans. John P. Leavey, Jr. (1962; Stony Brook: Nicolas Hays, 1978), pp. 102–3.

7. Jacques Derrida, *Positions*, trans. Alan Bass (1972; Chicago: University of Chicago Press, 1981), pp. 42–43.

8. Jacques Derrida, *Dissemination*, trans. Barbara Johnson (1972; Chicago: University of Chicago Press, 1981), p. 238. Hereafter cited in the text as *Dis*.

9. Jacques Derrida, "Signéponge," *Digraphe* 8 (1976): 33.

10. Martin Heidegger, "Moira," in *Early Greek Thinking*, trans. David F. Krell and Frank A. Capuzzi (New York: Harper and Row, 1975), pp. 79–101.

11. Jacques Derrida, "Form and Meaning: A Note on the Phenomenology of Language," in *Speech and Phenomena*, trans. David B. Allison (Evanston: Northwestern University Press, 1973), p. 108.

12. I am thinking of, for example, "D'un ton apocalyptique adopté naguère en philosophie," in *Les Fins de l'homme; a partir du travail de Jacques Derrida* (Paris: Galilée, 1981); and "Nietzsche's Otobiography" (a paper delivered at Yale, Emory, and the University of Florida, Spring 1982).

13. Roman Jakobson, "Poetry of Grammar, Grammar of Poetry," *Lingua* 21 (1968): 605.

14. Wendy Steiner, *Exact Resemblance to Exact Resemblance: The Literary Portraiture of Gertrude Stein* (New Haven: Yale University Press, 1978), p. 150.

15. Willy Rotzler, *Constructive Concepts* (New York: Rizzoli, 1977), p. 148. Cf. also J. O. Robinson, *The Psychology of Visual Illusion* (London: Hutchinson, 1972), and R. L. Gregory, *The Intelligent Eye* (New York: McGraw-Hill, 1970).

16. Jacques Derrida, *La Carte postale: de Socrate à Freud et au-delà* (Paris: Aubier-Flammarion, 1980), p. 124.

17. Walter J. Ong, *The Presence of the Word* (Minneapolis: University of Minnesota Press, 1981), p. 46.

18. Jacques Derrida, *Glas* (Paris: Galilée, 1974), p. 233.

19. E. H. Gombrich, *The Sense of Order: A Study in the Psychology of Decorative Art* (Ithaca: Cornell University Press, 1979), p. 251.

20. Stanley Rosen, *The Limits of Analysis* (New York: Basic Books, 1980), pp. 41–51.

21. Jacques Derrida, "The Law of Genre," trans. Avital Ronell, *Glyph* 7 (Baltimore: Johns Hopkins University Press, 1980), p. 206. Hereafter cited in the text as *LG*.

22. See Peter Timothy Saunders, *An Introduction to Catastrophe Theory* (Cambridge: Cambridge University Press, 1980).

23. Saunders, *Catastrophe*, p. 1.

24. Jacques Derrida, *La Vérité en peinture* (Paris: Flammarion, 1978), p. 43.

25. Jacques Derrida, *The Archaeology of the Frivolous: Reading Condillac*, trans. John P. Leavey, Jr. (1973; Pittsburgh: Duquesne University Press, 1980), pp. 73, 82, 118, 128, 133.

26. Jacques Derrida, "Living On: Border lines," trans. James Hulbert, *Decon-*

struction and Criticism, ed. Harold Bloom et al. (New York: Seabury Press, 1979), pp. 99–100. Hereafter cited in the text as *LO*.

27. C. C. T. Baker, *Introduction to Mathematics* (New York: Arco, 1974), pp. 51–2. Cf. Patrick Hughes and George Brecht, *Vicious Circles and Infinity* (New York: Penguin, 1979).

28. James H. Bunn, *The Dimensionality of Signs, Tools, and Models* (Bloomington: Indiana University Press, 1981), p. 83.

29. Lucien Dällenbach, *Le Récit spéculaire: essai sur la mise en abyme* (Paris: Seuil, 1977), pp. 17–18.

30. Ibid., pp. 18–19. Derrida discusses *Arnolfini and His Bride* in "Restitutions" (*La Vérité en peinture*, p. 399), and the notion of "mise en abyme" in general. He indicates his reservations here and also in "Speculations—on 'Freud'," trans. Ian McLeod, *Oxford Literary Review* 3, (1978): 78–97.

31. Angus Fletcher, *Allegory:* The Theory of a Symbolic Mode (Ithaca: Cornell University Press, 1964), p. 128.

32. Ibid., pp. 112–13.

33. "Law of the *parergon* which comprises everything without understanding and perverts all relations of the part to the whole" (*Vérité en peinture*, p. 392).

34. Jacques Derrida, "Structure, Sign, and Play in the Discourse of the Human Sciences," in *The Structuralist Controversy*, eds. Richard Macksey and Eugenio Donato (Baltimore: Johns Hopkins University Press, 1972), p. 260.

35. James Ogilvy, *Many Dimensional Man* (New York: Oxford, 1977), pp. 46–47.

36. Gombrich, *The Sense of Order*, p. 134.

37. Jacques Derrida, "Restitutions of Truth to Size," trans. John P. Leavey, Jr., *Research in Phenomenology* 8 (1978): 25.

38. Jacques Derrida, "Limited Inc," *Glyph* 2 (Baltimore: Johns Hopkins University Press, 1977), pp. 188–89.

39. A. A. Roback, *Destiny and Motivation in Language: Studies in Psycholinquistics and Glossodynamics* (Cambridge, Mass.: Sci-Art, 1954), p. 179. The title of this book could serve as a subtitle for much of Derrida's work, beginning with *Glas*. But whereas Roback's attitude is noncommittal, Derrida's tone is parodic, at least when he is working with this topic.

40. Roback, *Destiny and Motivation*, p. 264.

41. Roback, *Destiny and Motivation*, p. 266.

42. René Thom, "Morphogenèse et imaginaire," *Circé* 8-9 (1978): 40. Cf. his *Structural Stability and Morphogenesis*, trans. D. H. Fowler (1972; Reading, Mass.: W. A. Benjamin, 1975).

STAGING
Mont Blanc

***

Herman Rapaport

How can we envisage a teleology of subjectivity which would have been
subjected to the critical examination of a Freudian archaeology? It would
be a progressive construction of the forms of the spirit, after the manner
of Hegel's *Phenomenology of the Spirit*, but one which, to a greater extent
than in Hegel, would unfold on the very terrain of the *regressive* analysis
of the forms of desire.

—PAUL RICOEUR, "The Question
of the Subject"

Mother of this unfathomable world!
Favor my solemn song, for I have loved
Thee ever, and thee only; I have watched
Thy shadow, and the darkness of thy steps,
And my heart ever gazes on the depth
Of thy deep mysteries.
—PERCY BYSSHE SHELLEY, *Alastor*

IT MAY HAVE OCCURRED to Percy Bysshe Shelley, who came to Plato with a
skeptical frame of mind,[1] that in the famous allegory of the cave which
Socrates relates (*Republic*, Book VII), the cave itself is a representation that
like the shadows cast on an interior wall by means of fire and human
shapes (both natural and fabricated), blocks us from seeing the Real. And
yet this very image of the cave is at the same time a forum, opening, or
stage by means of which Socrates can inspire Glaucon with the knowledge
of the pure forms, of that reality men are bound never to see. Only by
means of the cave image, itself nothing more than the kind of prop held up
by the men who walk past the wall, a prop carried here by the philosopher
before his pupil, can Glaucon respond to the eager instructor, "All this I
see." So much, then, depends upon a stage prop, upon the theatricaliza-
tion of philosophy. Socrates holds up an image, illuminates it by allegorical extension, and on a wall or mental screen a shadow is perceived—the
doctrine of pure forms. But what is most interesting is the way a prop

such as the cave image can suddenly turn into a stage, how an image, itself framed, can suddenly stage itself as stage and in that way absent itself or disappear from the viewer's consciousness as image, object, or prop. We recall that Socrates begins the allegory by saying to Glaucon, "Picture men dwelling in a sort of subterranean cavern. . . ." The word "picture" frames, stages, or encloses the image of the cavern, but the cavern will in turn stage or frame other images. What is most peculiar is that Glaucon never calls attention to this bit of Platonic *trompe l'oeil*, this way of making image into theater, prop into stage.

I started out by asking whether Shelley might not have been aware of this Socratic way of seeing, because, as James Rieger notices in *The Mutiny Within*, when Shelley tries to view an object as it really is, "he either screens the dominant image or takes its measure against a frame in the foreground. Most characteristically, he establishes perspective by looking *through* one thing *towards* another. . . ."[2] Thus in a letter to Thomas Peacock dated March 23, 1819, Shelley writes, "I see the radiant Orion through the mighty columns of the temple of Concord."[3] Like the cave image in *The Republic*, the temple functions as an image that we look at, but one that also disappears as it turns into a frame or stage for an image that appears as if within, the radiant Orion. Of course, there is nothing remarkable about seeing one thing through another in this way. But when one encounters the following passage from *Alastor*, this mode of seeing may seem less familiar:

> The oak,
> Expanding its immense and knotty arms,
> Embraces the light beech. The pyramids
> Of the tall cedar overarching, frame
> Most solemn domes within, and far below,
> Like clouds suspended in an emerald sky,
> The ash and the acacia floating hang
> Tremulous and pale. . . .
>
> [11. 431–38]

In such passages it is clear that images function as frames, and sometimes for images that could be said to be specular doubles. The oak embraces the beech tree, but in doing so reminds us that it is a tree framed by a tree that we are seeing here, a tree at once embedded and embedding, imaged forth and framing. Similarly the "pyramids" are said to "frame" the "domes within." Then one notices by way of comparison between the clouds suspended or framed in an emerald sky and the ash and acacia, that it is through the emerald sky that these tree-clouds are to be envisioned. It is not so much that Shelley depends upon the strict correspondence or analogy between images in order to make comparisons, something that is rather obvious in the above example, but that he *stages* his similes by way of making one image the frame, set, or screen upon or through which another image is viewed. The effect of such vision is phantasmic, a point Shelley himself recognizes when he writes towards the end of *Alastor*, in an address to the Poet who has died on his stone slab: "Thou canst no longer know or love the shapes / Of this phantasmal scene. . ." (ll. 696–697). Indeed, it is Shelley's staging mechanism, what amounts to a poetic mode of ocular embedding and dissolution of imagery, that carries us from the natural to the supernatural, the canny to the uncanny, the image to the phantom or spirit.

Such phantomization is already there in the cavern mentioned by Socrates, on the wall which is screening the shadows. And one might argue that the "force" of Socrates' allegory depends upon an image that knows how to fade into a frame, that the phantomization Glaucon "sees" within the image of the cave is the consequence of the staging enacted by the image, a staging that counts on the dissolution or phantomization of an image in order that we can see through it, or, to put it another way, in order that we can see upon the image's faded surface something else. The image fades, then, and in doing so makes us "forget" this image even as we use it to see something else. What makes the image phantasmic (or powerful) in its effect upon us is the fact that even if we can "forget" it, the image's impression remains. In fact, this impression "houses" or "supports" ("props" is the best term, perhaps) the gaze of the reader, the inquiry of the pupil, Glaucon.

We are close to an observation of Geoffrey Hartman's: images serve "to stabilize a phantasm or to frame a fantasy."[4] This is implicit in Sigmund Freud's work when he documents what happens to a small child who has been exposed on repeated occasions to the picture of a wolf, an image that can be seen with or through like a kind of optic glass and thus can frame what will become a traumatic fantasy, a nightmare about six or seven wolves in a tree.

> I dreamt that it was night and that I was lying in my bed. (My bed stood with its foot towards the window; in front of the window there was a row of walnut trees. I know it was winter when I had the dream and night-time.) Suddenly the window opened of its own accord, and I was terrified to see that some white wolves were sitting in the big walnut tree in front of the window. There were six or seven of them.[5]

Here the image of the wolf has been phantomized, has faded out, and frames or stages this dream. Although the wolf image has disappeared in its original form, its effect or impression energizes the dream, and it is recollected or repeated six or seven times within the image's little "production," a "production" which is itself seen through a window frame and all

because the window can "open of its own accord." Apparently the frame
of the wolf image has been metaphorically displaced, figured forth as the
window in the dream. Jacques Lacan calls this frame the "cadre du désir,"
or "frame of desire," which, as Catherine Clément notes, is symbolized as
the "the half-open door, the skylight in the night, the open window."[6]
From a Lacanian perspective the phantasm is the frame, is this dissolution
of the image into a staging of desire.

If the allegory of the cave allows us to consider what would happen to a
person who could leave the subterranean world, the opening of a window
in the Wolf-Man's dream allows the dreamer to open his eyes. It is during
analysis that the Wolf-Man recognizes what the open window means,
"My eyes suddenly opened," and Freud argues that what the eyes opened
upon was a memory of something very real, a primal scene in which coitus
is performed *a tergo*, much like wolves. But if the Wolf-Man "sees" his
parents making love, he does so only through an image, the phantasmic
picture of wolves in a tree. We have noticed that initially the image of a
wolf seen in consciousness turns into a stage in sleep, what Lacan calls the
"frame of desire." And through this frame we see wolves in a tree. However, where there was an easy and unpremeditated "dissolve" of the initial
wolf image, we discover that there is something more like blockage with
the image of the wolves in the tree. Let us say for convenience that what
the window in the dream opens upon is much like a theatrical backdrop
that is extraordinary in one major respect: it is also a theatrical curtain that
opens onto another stage behind it, onto an obscene and traumatic spectacle that has already been seen and remembered. Thanks to repression, the
moment the window opens of its own accord, like a door in a haunted
house, this curtain drops on a most intriguing but frightening spectacle in
a bedroom. But if the spectacle behind the backdrop is traumatic, so is the
image of the protective curtain or backdrop, and thus because of a double
bind the dreamer can be said to lose himself in the arras, cannot tell which
side of the curtains he is on and cannot differentiate clearly what he has
seen behind the curtains from what he sees designed on them. What
makes staging pathological in the Wolf-Man's dream is this refusal of an
image to be either fully opaque or fully transparent, the refusal to fade and
the inability to block yet another *mise en scène*. It is such pathological
staging that makes obsession possible.

In the Wolf-Man's dream, then, there is a succession of stagings, and
not only do images frame scenes, but a scene also draws its own curtains,
calls for the rematerialization of an image to block (however ineffectively)
the dreamer's gaze. Not only fading, then, but blocking or mediation of
the image or prop is needed for the kind of theater one encounters in
psychoanalysis. This is true as well for philosophy and literature, for
there too one posits scenes or acts beyond comprehension that must be

staged in order that the unrepresentable be represented. Thus Socrates' allegory of the cave attempts to figure forth an Ur-scene of pure forms that the philosopher, not unlike the Wolf-Man, has already "seen" or witnessed. And yet with what curtains or backdrops must the philosopher struggle? Similarly, Shelley's poet in *Alastor* searches for his "vision," which is inaccessible, though representable in terms of a quest in which there occur numerous stagings similar to the kind we encounter in the Wolf-Man's phantasmic experience.

In "Shelley Disfigured," Paul de Man observes that in *The Triumph of Life* the "polarities of waking and sleeping (or remembering or forgetting) are curiously scrambled . . . with those of past and present, of the imagined and the real, of knowing and not knowing." De Man could have been addressing Freud's case history of the Wolf-Man and in particular the dream we have briefly considered. But de Man is reading these lines which Shelley has Rousseau speak:

> "So sweet and deep is the oblivious spell;
> And whether life had been before that sleep
> The Heaven which I imagine, or a Hell
>
> "Like this harsh world in which I wake to sleep,
> I know not."
>
> [ll. 332–335]

"We cannot tell," de Man writes, "the difference between sameness and difference, and this inability to know takes on the form of a pseudo-knowledge which is called forgetting."[7] This is exactly the kind of pseudo-knowledge the Wolf-Man has when he cannot tell which side of the curtain he is on, the kind of knowledge revealed by images which make themselves into undecidable partitions. What is of interest may not be so much the issue of "disfiguration," what de Man calls the loss or fading of a face, the erasure of self-knowledge, but how such a forgetting, manifested in terms of the difficulty of discriminating difference and sameness, gives rise to the kind of staging or framing of desire that we noted in *The Republic*, *Alastor*, and *History of an Infantile Neurosis*. It is in terms of Shelley's fascination with images possessing a radical ambivalence with regard to the difference of life and death that such a framing or staging not only occurs, but in terms of a certain dis-placement, an ob-scene that concerns a set or scene that cannot be directly viewed. Moreover, what will interest us is how the occlusion of an image is invested with a strong regressive tendency which depends upon props and stages, a theatrical tendency that, as a matter of fact, has everything to do with what Freud called the *anaclitic*, a leaning of the drive, which will concern itself with a particular mountain in Haute-Savoie.

Perhaps what de Man calls "disfiguration" has been conceptualized earlier and more theatrically as the "ob-scene" by Philippe Lacoue-Labarthe, who in "La scène est primitive" writes that, "like the female or maternal genitals, death cannot present itself as such, or as Lyotard would say, 'in person.'"[8] Death has to be displaced, shifted, dissimulated, that is to say, forgotten, in order to be represented or viewed, put on stage or "seen" as "scene" and thereby identified or mimicked. Like coitus *a tergo* in the case history of the Wolf-Man or like Medusa's "genitals," death is viewed obscene or away from its proper and unmediated spectacle. The hyphenating of the word ob-scene calls attention to the prefix "ob" (in Latin a preposition meaning in front of, in view of, towards, but also against), which in relation to the "scene" signifies a displacement or removal of a spectacle from the viewer, a distance placed between subject and object. The ob-scene is the scene before the scene or a scene against a scene. To quote Lacoue-Labarthe:

> Disons, s'il est permis de jouer sur une étymologie "populaire," que le mort est *ob-scène*. Ce que sait du moins Freud, c'est que la mort "ne se regarde pas en face" et que l'art (comme la religion) a ce privilège d'être le commencement de la représentation économique—c'est-à-dire de la représentation libidinale. La mort n'apparaît jamais comme telle, elle est strictement *imprésentable*,—c'est l'imprésentable meme. . . .

> If it is permissible to play on a "popular" etymology, we might say that death is *ob-scene*. At the very least, Freud is convinced death "cannot be looked in the face" and that art (like religion) has the privilege of being the beginning of economic representation—that is, of libidinal representation. Death never appears as such, it is in the strict sense unrepresentable, or the unrepresentable itself. . . .[9]

Thus any scene representing death merely places itself in front of a scene which properly speaking cannot present itself. To talk about the obscene is to recognize the distance or difference that displaces—in terms of the Freudian sense of staging, dreaming, fantasizing—the *mise en scène* that takes place in front of the primary scene, whatever that may or can be. Such a *mise en scène* is deferred, but is also subject to a disfiguration, to the pseudo-knowledge produced by a scene which is but a phantom proxy. Like the backdrop of the wolves sighted by the neurotic child, the *mise en scène* before death is itself subject to phantomization, erasure, or disfiguration, thus allowing itself to open or stage a theatrical production which again only veils that which cannot be represented. Even in the case of the Wolf-Man the curtain opens in this way, stages the set for the child's gaze by way of recalling a memory of coitus *a tergo*, a primal scene that is viewed as ob-scene in so many senses. The backdrop (or curtain, as I have also called it) is at once opened and closed, presented and removed, known and not known, the same as and different than. It is what Lacoue-Labarthe calls the *theatrum analyticum*, and nowhere is it put to better advantage than in the drama of the romantic period, as Lacoue-Labarthe makes evident in "The Caesura of the Speculative," an essay on Hölderlin's adaptation of *The Antigone*.[10]

Certainly Shelley's *Prometheus Unbound* is an excellent example of a romantic staging of a death that never manifests itself as anything other than metaphor, vision, tableau, or theater: as a man chained to a precipice, who is suffering from having his heart devoured by a vulture, a creature that never gets its pound of flesh no matter how much is taken from the heart of man, as if to insist that death is pure chimera, a limit that displays itself only in terms of delay, deferral, distancing, eternal return, or, to borrow a phrase from Maurice Blanchot that has been much on the mind of Jacques Derrida, an *arrêt de mort*, a "death sentence" and/or "stay of execution."[11] Indeed, much of *Prometheus Unbound* is but the dramatization of this "stay" in which the difference between being put to death and being kept alive is "curiously scrambled," undecidable. Moreover, this "stay" or refusal, what Blanchot would call a suspension, is marked all the while by discourse, by the temporality of language. And perhaps Shelley, however indirectly, may be asking in his poetry a question akin to that asked by Blanchot and Derrida: how does one give the sentence of death to language, to the sentence that pronounces sentence itself? What would constitute a "death sentence" in prose, poetry, or drama, for that matter?[12]

Certainly in *Promotheus Unbound* the hero is at the *arrêt de mort*, is, like the figures in Blanchot's fiction, suspended in an "écriture du désastre," but he is also at the *arête de mort*, or ridge or arris of death, while still discoursing, having an *entretien infini*, and right there on the *arête* or peak, that border line (to recall the pertinent essay of Derrida's) or precipice of words.[13] This ridge, we recall, is not just an image, but a kind of stage, a vertical platform upon which the drama is played out, upon which the hero, Prometheus, is suspended and his "death" re-presented (repeated) ob-scene. This theatrical rock even speaks as the character Earth in *Prometheus Unbound* and might lead us to consider at length the triumph of Promethean life were it not that something off-stage or ob-scene, not unlike a faded memory, catches our attention: an "other" *arête*, *Mont Blanc*.

It is as if *Mont Blanc* anticipates that which takes place in *Prometheus Unbound*, as if desire relays itself back and forth between these two literary sites, each already displaced or displacements, as if one *topos* is to be seen through another *topos*. In this sense, art is not only chiastic, but also a recurring nightmare about that which cannot appear to us "in person," but only as phantom. The *arête* (ridge, mount, peak, spine, cross-point) may come to be seen as a perpetual shifting, an eternal return to the same as different, an obsession where desire is fixed or pinned down, and precisely there where one has a suspension, not unlike the suspensions in the Wolf-Man's dream. Here the ability to distinguish between identity and difference is lost, for slippage occurs, what de Man in "Shelley Disfigured" calls "forgetting," characterized by this suspension between two sites or scenes (de Man is considering *The Triumph of Life*) at once similar and dissimilar, what I have called earlier a *pathology of staging*. It is a question now of where we are, of a representation that cannot decide a suspended relation and which proposes itself as a repetition (a re-presentation) and therefore as both remembered and not remembered, as present and as lost, visible and invisible. Serge Leclaire in *Démasquer le réel* points out most convincingly that at such points in which the laws of contradiction are suspended, fixated representations make their symptomatic appearances.[14] Suddenly images of power raise themselves up, erect themselves in the thoughts of the patients, and the metapsychological or economic reasons for this are clear: the subject does not want to lose his energy, to break into pieces, but to conserve himself in the monolithic *aporia* of an axis or crossing point that is endlessly forestalled in the undecidable suspension of an *arrêt de mort* whose herald or emblem is the *arête de mort*. Recall that on the Promethean rock desire is chained for the purposes of eternal sparagmos, repetition, and again, in Blanchot's terms, condemned to speak an infinitely drawn out dialogue, one that is thoroughly fatigued and ends in the full *arrêt*, the sentence that condemns and gives reprieve at the same time, an *arrêt* that, however much it cuts into the sufferer and dismembers him, also keeps him whole and intact, preserves him on the rock eternally.

In *Mont Blanc* the *arrêt / arête* (the suspension and the re-presentation that marks it) figures itself forth most blatantly as a signifier of power, and what is most interesting about this particular representation of the *arrêt* is that it is the major prop or stage upon which the discourse of the poem rests, the site or theater for another infinite discourse of subject and object, a dialogue that is very problematic in Shelley, as critics like Earl Wasserman demonstrate. Wasserman's argument, that for Shelley "the ultimate doctrine of the Intellectual Philosophy is that reality is an undifferentiated unity, neither thought nor thing, and yet both," fits in perfectly with Blanchot's and Derrida's thoughts on an *entretien* that breaks limits, border lines, horizons, that suspends philosophical speculation and throws man into the condition of living out a death sentence.[15] In short, what Shelley believes, according to Wasserman, is that life (thought, discourse, consciousness) and things (matter) cannot be differentiated or confused either. Rather, this opposition has to be put out of order.

Wasserman agrees as well that "more precisely, the poem *Mont Blanc* is not lyric but dramatic."[16] And therefore the poem is akin to *Prometheus*, a connection we have already discovered for ourselves. In any case, let us assume for the sake of argument that in *Mont Blanc* the Earth speaks, the stage has its say, that the mountain or landscape does not simply support by means of metaphor or allegory a developing consciousness, of what amounts to philosophical speculation on sublime sights, but serves as a stage that raises or props up the whole problematic of the Promethean suspension between life and death, raises the *arête de mort* as an *arrêt de mort*, as "stays" in a supporting or propping of the impossibility of either life or death. Then it remains for us to investigate how this stage props, or, for our purposes, what the notion of the "prop" signifies for our analysis.

So I will just say one more thing about Shelley's poem: the mountain, a signifier of power and desire, is anaclitic. That is, Mont Blanc is an *étayage*, a supporting or propping up (*étager?*) whose function is not unlike the mother's breast as Sigmund Freud and Melanie Klein discuss it in their work on infant sexuality. That is to say, the mountain is not phallic but maternal, like Earth in *Prometheus Unbound*, for it is the Object (Lacan's *objet a*) on which the drives lean. And this is really one of the purposes of my paper: to "decide" or "stage" this perhaps silly or frivolous question, put to us by Freudians of an earlier age, as it were, of whether the mount is phallic or maternal; that is, to speak in this gap of gender, in the obscene and on the *arête* where the drives attach themselves, and, it must be admitted, save themselves.

Jean Laplanche explains this notion of the leaning of the drive when he writes,

> Ainsi le terme d'étayage a été compris . . . comme un appui sur *l'objet*, et finalement un *appui sur la mère*. . . . Ce qui est décrit par Freud c'est un phénomène d'appui *de la pulsion*, le fait que la sexualité naissante s'étaye sur un autre processus à la fois similaire et profondément divergent: la pulsion sexuelle s'étaye sur une fonction non sexuelle, vitale ou, comme le formule Freud en des termes qui défient tout autre commentaire, sur une "fonction corporelle essentielle à la vie."

> Thus the term *propping* has been understood . . . as a leaning on the *object*, and ultimately a *leaning on the mother*. . . . The phenomenon Freud describes is a leaning *of the drive*, the fact that emergent sexuality attaches itself to and is propped upon another process which is both similar and profoundly divergent: the sexual is propped upon a nonsexual, vital function or, as Freud formulates it in terms which defy all additional commentary, upon a "bodily function essential to life."[17]

Sexuality is ob-scene to the degree that it is a function that imitates another function which is inherently nonsexual. Sexuality leans on the child's experience of sucking the breast for the purposes of nourishment, or, as Freud writes, "sexual activity attaches itself to functions serving the purposes of self-preservation."[18] But if sexuality leans on a function, it is also propped up by an object, the mother's breast. The sexual drive asserts itself when the breast is removed, when the breast can only serve as a symbolic support, as an image or part-object, when the breast menaces with its absence, its after-image of the having-been-present. That is to say, the breast makes an appearance, sooner or later, as a representation, and this image is, like Plato's famous cavern, subject to fading or disfiguration. This occurs because the sexual drive defines itself most strongly when the breast reveals itself as the peak of deprivation *(arête de mort)* or the withholding of nourishment, but also as the *arrêt* of deprivation, the pleasures of feeding, the horizon of fulfilled desire, as potentially present even when absent. In this sense the breast is a prop not unlike a stage prop which serves as a stage set for sexuality, which becomes a theater where the drives can lean, where such drives can play out their scenario using whatever props are at hand. If this sounds odd, recall that a well-known American psychoanalyst, Bertram Lewin, has advanced precisely this argument when he claims in "Sleep, the Mouth, and the Dream Screen" that the mother's breast is introjected by the infant as a "dream screen" upon which or through which other images present themselves, for particularly after feeding the child sleeps and stages whatever scenes he wishes on the imaginary maternal screen. What is significant for Lewin is that adults "regress" at night when they sleep by re-envisioning this screen, by propping or leaning the drives on this phantasmic sheet or curtain which, as Lewin's patients say, "unfolds" or simply "unrolls." Such an *étayage* is nothing less than a compulsive re-staging of a primordial relationship with what Lacan calls the *objet a* (the Mother). And this *objet a*, according to Lacan, comprises the locus of "the gaze" *(le regard)*.[19]

Lacan, not unlike Melanie Klein and Bertram Lewin, argues that the breast is by no means unambiguously cathected; it is not just pleasurable, but menacing or threatening as well. Indeed, the breast is an *arrêt de mort*, as far as the child knows, a sentence pronouncing death (the end of nourishment, peace, bliss, attachment) and a stay or reprieve from being cut off from the pleasures of the mother. And it is this *arrê de mort* that the obsessional craves, as Serge Leclaire makes explicit when he observes that obsessives like Philo (a patient so dubbed by Leclaire) see themselves as the proper objects for their mother's desire, as "the chosen," but that such an alliance (Eros) is paid for by a heavy mortgage: the interdict of incest, which articulates itself as a death sentence. In the case history of another patient, Jerome, Leclaire writes (and here note the *arrêt* in the original text):

> . . . ce qui frappe le plus Jérôme, c'est la formule que prononce *le juge lorsqu'il rend son arrêt*: ". . . est condamné à être pendu par le cou, jusqu'à ce que mort s'ensuive."

> Eh bien, pour moi, ajoute-t-il, c'est comme si l'on m'avait dit un jour:
> *Tu vivras jusqu'a ce que mort s'ensuive.*

> . . . but most striking to Jerome are *the judge's words when he passes sentence:*
> ". . . is condemned to be hanged by the neck until dead."
> "Well," Jerome says, "for me it is as though someone had said to me one day,
> '*You will live until dead*'."[20]

An obsessional patient has articulated yet once more an *arrêt de mort* whose purposes is to evade the differences between life and death, to forestall the deciding of this antinomy. And this forestalling is nothing less than a regressive activity, as we will see shortly. At the axis or arris of the cross between life and death is implanted the neurotic's own tombstone, *arête*, or mound. Am I dead or alive? Jerome cannot answer, exactly, but he has half-remembered dreams to show as compensation for what he does not know, faded dreams about mummies, corpses, and tombs. What is problematic, implies Leclaire, is that Jerome has constructed a theater in which he cannot decide whether to play Oedipus or the Sphinx, in which one figure is invaginated in the other, and thus Jerome stars in both roles at the same time. As Oedipus, Jerome questions himself from within the figure of the Sphinx; which is to say, each character becomes a prop for the other. But it is in the case histories of Philo and Ange Duroc (The Angel of the Rock) that we learn who this Sphinx really is, and that is no one else than mama. It is she for whom Philo longs, and she whom the Angel of the Rock wishes to save (for himself) even at the expense of his own sexuality. At the crossroads, then, one encounters an *arrêt de mort* and clings to it for dear life, or should we say, to mother's breast? Clearly we are now far beyond the pleasure principle, though by no means within the vicinity of the death drive. Rather, we are in a kind of eddy, a suspension in which the drives circulate without going anywhere.

And yet, for all that, there is direction in the sense of a regression. Thus it is to *Mont Blanc* that I must return, particularly to the lines introducing part three.

> Some say that gleams of a remoter world
> Visit the soul in sleep,—that death is slumber,
> And that its shapes the busy thoughts outnumber
> Of those who wake and live.—I look on high;
> Has some unknown omnipotence unfurled
> The veil of life and death? or do I lie
> In dream, and does the mightier world of sleep
> Spread far around and inaccessibly
> Its circles? For the very spirit fails,
> Driven like a homeless cloud from steep to steep

That vanishes among the viewless gales!
Far, far above, piercing the infinite sky,
Mont Blanc appears,—still, snowy, and serene—

[ll.49–61]

The "veil of life and death" is not unlike a screen upon which the pleasure and death drives represent themselves, impossibly, undecidably, liminally. It is that stage or platform (veil as mount) between sleeping and waking, a kind of mystic writing pad articulated at the preconscious level, somewhere between looking and dreaming, gazing and recollecting images involuntarily, either in the tranquility of sleep or the astonishment, so overdetermined, of serene wakefulness, like the Wolf-Man's casual glance. What else is this mount, this blank mountain, but the maternal screen upon which the child individuates with pleasure, and precisely because the child recognizes the power of the breast, this transitional object or prop figured forth in regions not wholly conscious? This *arrêt de mort*, this sentence that prolongs life and commands death, produces pleasure and pain, relief and anguish, presence and lack, this prop, so still, snowy, and serene, is, like woman, veiled. It is a part of woman that needs to be unveiled by some "unknown omnipotence," the child's desire to "see," to "stage" his (Shelley's?) desire, to situate an Eros/Thanatos whose difference will not simply cleave in two up there on the invisible white heights.

From this perspective, *Mont Blanc* is situated in terms of a *theatrum analyticum* where the drives are featured as stage props, in terms of the stage itself, which is its own prop or "set" of props, or, to put it another way, a base that is its own superstructure, a Hegelian stage of *Aufhebung*: Ca as SA *(Savoir Absolu)*.[21] That is, with Hegel, but also with Shelley, Plato, and Freud in mind, we find it naïve to consider Mont Blanc merely an image, for this mountain is precisely the kind of prop that frames or stages desire, the phantomizing image that turns into a forum and frames as well as stabilizes fantasy. Mont Blanc is the dream screen, to recall Lewin, which facilitates a passive and regressive return to the maternal, a return that is obsessively concerned with an undecidable problem: this breast is not merely a site for past pleasure but constitutes a crossroads, an *arrêt/arête de mort*.

In *Mont Blanc* the alliance of the maternal with death or threat reveals itself as "the naked countenance of earth / On which I gaze," where the phallic mother presents herself, is under-erection (in every sense), for there the "glaciers creep / Like snakes that watch their prey. . ." (ll. 100–101). Perhaps it is all a defense, this apotropaic leaning of the drive, and from what else, a Lacanian like Leclaire would insist, but that which the phallic mother veils, the sight which must, according to Freud in "Medusa's Head," turn the spectator to stone (to an Ange Duroc): the "Dizzy Ravine"?

> . . . and when I gaze on thee
> I seem as in a trance sublime and strange
> To muse on my own separate fantasy,
> My one, my human mind, which passively
> Now renders and receives fast influencings . . .
>
> [ll. 34–38]

Before this ravine the mind is rendered passive; it cannot engage the landscape in any other manner but in finding fantastic substitutions for it. Thus the mind produces "wild thoughts" which rest

> In the still cave of the witch Poesy,
> Seeking among the shadows that pass by—
> Ghosts of all things that are—some shade of thee,
> Some phantom, some faint image; till the breast
> From which they fled recalls them, thou art there!
>
> [ll. 44–48]

In the cave the mind seeks some faint image of the ravine, as if the ravine itself were perpetually ob-scene (as ob-scene as the mount), and precisely for the reason that this ravine is vaginal, is that part of woman that has to be retracted, derealized, dissimulated, faded, even as the mind contemplates from within the security of this trench, this witch or Sphinx, poesy, this "still cave." As Lacoue-Labarthe might say, "La scène est primitive." Moreover, it is but a repetition of what Shelley himself calls the "breast" from which the thoughts fled: the *arête/arrêt de mort*. Mont Blanc is what Shelley himself calls "A city of death . . . Yet not a city" [ll. 106–8]. It is some kind of horrible power that cannot be precisely specified, looked at, determined, and yet, for all that, this mountain is the screen upon which the poet individuates with pleasure. And because he recognizes the power of this mount as a prop for fantasy, a dwelling place for the imagination, he comes to accept it as the rightful stage, spatially and temporally, to situate his desire, as the *theatrum analyticum* in which even philosophies of the mind can be figured forth.

If more conservative readers of *Mont Blanc* have followed with great care Shelley's theory of mind, particularly its development by stages, it has been my wish to posit an archaeology of the libido upon which any such theory is inevitably couched, an archaeology of figural attachments and stagings whose regressive force we have recognized and surveyed without disdain. Indeed, such a return to the primitive is all a matter of saving a text from its own death sentence, of what Derrida calls *sur-vivre* in relation to *The Triumph of Life*, a survival that has everything to do with a coming into one's own that is peculiarly attached to a secure loan, fastened onto a maternal bond that is safely kept in its vault or tomb, off-stage or ob-scene. If the poet as obsessive builds his monoliths at the crossroads, places his *arête* on the *arrêt de mort*, it is not only to ward off death by means of an ambivalent forestalling, but to build a monument to mother, to worship the virgin and child, or *le très haut* (the All High), to recall Blanchot once more. It is this debt to the past, always symbolic, that underwrites Shelley's poem and engenders what we will call a "passive tropology," something Leclaire's obsessive patients would find quite familiar. Of such passivity Blanchot writes,

> La passivité est sans mesure: c'est qu'elle déborde l'être, l'être à bout d'être—la passivité d'un passé révolu qui n'a jamais été: le désastre entendu, sous-entendu non pas comme un événement du passé, mais comme le passé immémorial *(Le Très-Haut)* qui revient en dispersant par le retour le temps présent où il serait vécu comme revenant.

> Passivity is without measure: it goes beyond being, being at its very limits—the passivity of a bygone past that never has been: the disaster understood, deeply felt, not as an event of the past, but as the immemorial past (the All High [*Le Très-Haut*]) which comes back while scattering by its own return the present time in which it would be experienced as coming back.[22]

Another title, *Le Très-Haut*, re-marks the most extreme suspension between life and death, is synonymous with *L'Arrêt de Mort*, but also refers to a man being propped up by a woman, raised by her, lifted in a *récit* to *La Plus Haute*. And can there be any doubt that *le très haut* in the passage above taken from *L'Écriture du désastre* is not in some sense already visible in *Mont Blanc*, in the Haute Savoie, and that it is not always already a passive monument which overcomes itself on the high road to self-consciousness in that curious return to an immemorial past that scatters the very present by which it can only return, that regressive and obsessional urge that it is the poet's fate to live over: *sur-vivre?*

NOTES

1. Shelley's skepticism with reference to Plato is most convincingly argued in C. E. Pulos, *The Deep Truth* (Lincoln: University of Nebraska Press, 1954).
2. James Rieger, *The Mutiny Within* (New York: Braziller, 1967), p. 92.
3. Cited in Rieger, p. 92.
4. Geoffrey Hartman, *Criticism in the Wilderness: The Study of Literature Today* (New Haven: Yale University Press, 1980), p. 26. The context of the quotation is as follows: "I have suggested that the image of Yeats's poem serves to stabilize a phantasm or to frame a fantasy. It is tempting to guess at an equation: the more image, the more fantasy."
5. Sigmund Freud, "History of an Infantile Neurosis," in *The Wolf-Man by the Wolf-Man*, ed. Muriel Gardiner (New York: Basic Books, 1971), p. 173.

6. Catherine Clément, *Le pouvoir des mots* (Paris: Mame, 1973), p. 100. My translation.

7. Paul de Man, "Shelley Disfigured," in *Deconstruction and Criticism*, ed. Harold Bloom et al. (New York: Seabury Press, 1979).

8. Philippe Lacoue-Labarthe, "La scène est primitive," *Le sujet de la philosophie* (Paris: Flammarion, 1979), p. 206; "Theatrum Analyticum," trans. Robert Vollrath and Samuel Weber, *Glyph* 2 (Baltimore: Johns Hopkins University Press, 1977), p. 135. Lacoue-Labarthe is referring to Lyotard's introduction, "Par delà la representation," to Anton Ehrenzweig's *The Hidden Order of Art* (Paris: Gallimard, 1974).

9. Lacoue-Labarthe, "La scène," p. 206; "Theatrum," p. 135.

10. Lacoue-Labarthe, "The Caesura of the Speculative," in *Glyph* 4 (Baltimore: Johns Hopkins University Press, 1978), pp. 57–85. In French, *Hölderlin: L'Antigone de Sophocle* (Paris: Christian Bourgois, 1978).

11. *L'arrêt de mort* (Paris: Gallimard, 1948) is a novel by Maurice Blanchot which plays on the *double entendre* of a sentence which condemns and reprieves.

12. This question is pursued in Jacques Derrida, "Living On: Border Lines," in *Deconstruction and Criticism*.

13. See Derrida, *Deconstruction and Criticism*, p. 109, for details on the prolonged play upon the words *arrêt/arête*. In his essay on Blanchot and Shelley, Derrida takes over Blanchot's play with *arête/arrête, arris/stay*. Here an extra or supplementary "r" is added in order to conflate a noun with a verb.

14. Serge Leclaire, *Démasquer le réel* (Paris: Seuil, 1971), pp. 180–86.

15. Earl Wasserman, *The Subtler Language* (Baltimore: Johns Hopkins University Press, 1959), p. 204.

16. Wasserman, *Language*, p. 208.

17. Jean Laplanche, *Vie et mort en psychanalyse* (Paris: Flammarion, 1970), pp. 30–31; *Life and Death in Psychoanalysis*, trans. Jeffrey Mehlman (Baltimore: Johns Hopkins University Press, 1976), p. 16.

18. Sigmund Freud, *Three Essays on the Theory of Sexuality* (New York: Harper, 1962), p. 48.

19. Bertram Lewin, "Sleep, the Mouth, and the Dream Screen," *Psychoanalytic Quarterly* 15 (1946); "Le sommeil, la bouche et l'écran du rêve," trans J.-B. Pontalis, *Nouvelle Revue de Psychanalyse* 5 (Spring 1972). Also see Lacan, *Le séminaire: livre XI* (Paris: Seuil, 1973), p. 65 ff.

20. Leclaire, *Démasquer*, p. 128; "Jerome, or Death in the Life of the Obsessional," trans. Stuart Schneiderman, in *Returning to Freud*, ed. Schneiderman (New Haven: Yale University Press, 1980), p. 99. I have altered the typography of the translation to confrom with the original text. I have also added italics; only the sentence "You will live until dead" is italicized in the original.

21. Let us just say the floor is tilting, that what is flat becomes suddenly upright: the horizontal becomes vertical. Thus the mountainous platforms in Shelley, those gigantic props which are their own stages, which collapse and erect.

22. Maurice Blanchot, *L'Écriture du désastre* (Paris: Gallimard, 1980), p. 34. Passage translated by my colleague Andrew McKenna, with revisions of my own.

I wish to thank Alexander Gelley for comments and Rene Hubert for inviting me to lecture on this material at the University of California at Irvine. I am grateful to Geoffrey Hartman for some extremely important suggestions and to Mark Krupnick for his editorial help. Lastly, "Staging: Mont Blanc" is to be read with its companion piece, published in *Enclitic* (Fall 1982), entitled "The Disarticulated Image: Gazing in Wonderland," in which I explore more fully the Lacanian dimensions of these issues.

A TRACE OF STYLE

***

Tom Conley

HOW TO READ—TODAY—DERRIDA?

"COMMENT LIRE—AUJOURD'HUI—WARBURTON?" That is how, a few years ago, Jacques Derrida began an essay under the name of "Scribble" on the British archaeologist's study of hieroglyphics. Its tone rings true to many of Derrida's beginnings. In this fragmentary incipit—whose dashes mark a musical cadence—below the title, a reader acquainted with Derrida will already hear the limpid measure of his phrases that have sounded the chime (or bell at the beginning of a seminar) of philosophy. His first words incise their topic. They cut right into it, they summarize what is entailed by displacing us into the risky adventure of beginning any piece of writing.

"I will speak, therefore, of a letter": also wrote Derrida in 1968, when he inaugurated the concept of *différance*, which has become almost a password in American avant-garde criticism. "—Of the first (letter), if we are to believe in the alphabet and most speculations ventured about it." The letter was A, aleph, the figure of the pyramid and the law of the letter, the triumph of order, the delta of symmetry and desire, the eye on the back of a dollar bill.[1] Thus began a decade of philosophical inquiry. "I will resound a letter, therefore I will have been." His Cartesian rewriting of the same kind of incipit tells us how pervasively Derrida has been erasing philosophy's last words in what many readers have taken to be another version of Apocalypse.

For already, at the outset of *La dissémination*, Derrida had written: "This, therefore, will not have been a book." Or only two years ago, under the title "D'un ton apocalyptique adopté naguère en philosophie": "I will speak, therefore, of an apocalyptic tone in philosophy." Every one of these lapidary statements calls into question what we ask his books to do for us. They play back the title under which they rest, and their repetition excludes the reader from them—whether by reiteration of headings or by the suspensive use of the future perfect tense that disallows any appropriation of content outside of his texts' hermetic form. The relation of the titles to the discourse effaces whatever truth we wish to cull from

74

them. In an essay on Valéry, Derrida began, "I—mark right off, a division in what might have appeared at the beginning." Once again, the dash barring the verb from the first-person singular, or the subject from the predicate, closes us from entry into the very text on which our eyes are gazing. Winking, the dash marks the time of his entire speculation. The volume of his book is either reduced—to a line—or displaced into a bizarre hieroglyph that forever eludes our grasp.

Sometimes the beginnings are notes sketched, seemingly in haste, to get around the question of beginning. They are lines that betray the doubt of a writer who lets words elide and erase the hand that wrote them in the first place. Thus the incipit of Derrida's text "White Mythology":

> Of Philosophy, rhetoric. Of a volume, nearly, more or less—to make here and now a flower, to extract it, to mount it, to leave it, rather, to mount it, to be born—turning itself away as if from itself, in volutes again, such a heavy flower—learning to cultivate, according to the calculation of a lapidary, patience. . . .²

No sooner does "White Mythology" begin than it turns to its own silence, in ellipsis.

These incipits really tell us how difficult it is to read Jacques Derrida. So physical, so visceral, and so hermetically musical are his styles that they defy any summary, paraphrase, or translation that would try to appropriate them. The paragraphs that follow will explore what they take to be the difficult beauty of Derrida's writing simply because of its enduring commitment to *style*. The bias of this kind of reading involves setting aside the issues of deconstruction and criticism and the precise consequences of Derrida's engagements in given philosophical traditions. Accomplished readings of Derrida from the standpoint of philosophy and criticism abound in English, as do useful prefaces by Derrida's translators. The aim of the present text, however, is not identification with Derrida by way of a close reading. It is, rather, to mark here and there some movements and turns of his style that may have evaded an American reader.

In the last five years or so, Derrida has been breaking away from the modes with which most of his Anglo-American readers have been familiar, at least in the wake of *Grammatology* and *Writing and Difference*. In moving away from a manner of writing called "deconstruction," Derrida has been working under the press of fiction, or with a legacy of texts that have no generic stability. Less and less can his recent compositions be identified as products of academic seminars, as are many of his essays in *Marges*, *Dissemination*, the Hegel column of *Glas*, and the chapters on Kant entitled "Parergon" in *La Vérité en peinture*. Neither are they entirely circumstantial, prepared for colloquia, as were "La Loi du genre" (delivered at a meeting in Strasbourg on the problem of genre), "Le T. i. t. r. i. e. r. (titre à préciser)" (about the freedom accorded him in respect to the title and topic of a lecture he was invited to deliver at Brussels in 1979), or his multilaterally oriented essays on Blanchot's *Folie du jour*.

On the other hand, the philosophical concerns of the pieces on Valerio Adami and Hans Titus-Carmel, which Derrida wrote for exhibitions at the Maeght Gallery and the Beaubourg, make it impossible to label this writing as art criticism. To complicate further the problem of genre, all we have to do is ask: what is *La Carte postale?* Is it the book adapted from a seminar Derrida gave on *Beyond the Pleasure Principle* in the spring of 1976? A notebook that took the form of an epistolary romance? A self-defeating job application (for a position or *poste* the author did not obtain at Paris-Nanterre)? A fortune-telling book about the post-modern condition? A piece of auto-analysis gained through its encounter with all the pedagogues named throughout the book? Neither can the text be reduced to an essay on the contamination of genres. But its uneasy status can permit us to sketch—because of the exactitude of its own dating—how and where Derrida's own work may have displaced itself.

In the fall of 1978, at the time of a journey in the United States and Canada which *La Carte postale* marks by a temporal skip from October 13 to "the first days in January 1979,"[3] Derrida presented a number of lectures in California that explained the enigmatic topic he entitled *Du droit à la littérature*. He argued that the historical origins of literature can be found in jurisprudence. Letters are born from the strictures of legal writing; they are an extension of it. They often parody and pervert it, and they may try to escape it, but they are always subject to its codes of formulation. A historian of literature might argue that nothing could be more obvious. Montaigne and Corneille were lawyers; Pascal had an extremely legal sense of debate in the *Provinciales;* Stendhal read the civil code every morning before proceeding to work on his novels; and Gide filled his fictions with magistrates and advocates of the devil. But Derrida's approach was not exactly historical, for he was aiming at showing how *every* scene of writing reenacts the terror of an encounter—real, failed, fantasied, denied, or disavowed—with the Law. To respect and betray the Law, to cut along the edge of the sword of Justice, writing must advance toward and retreat from issues it is forced to espouse. It accepts, rejects, summarizes, and filibusters. To stay alive, it prolongs debate in a graphic rhetoric of chicanery and dissimulation. Because it is composed under the sign of Terror, its doubts, hesitations, and its varied tones become the styles of literature. The sight of the rapier of Justicia, the feel of the wedge of the guillotine cutting along our necks—these turn the discourse of Law into obscure, nightmarish scenes of fiction. What had been an art of copying, of calligraphy, gives way to a momentary absence of script. Displaced, the graphic language of Justice evokes death-drives and figures of monstrosity that play out dialogues for the sake of the writer's will to stay alive.

Here, according to Derrida, began the intransitive activity that we know today as *écriture*. As soon as it was born, it became an institution and was delivered from night back into day and thereby was once again lost from view. On more than one occasion, Derrida has said that in our time, more than ever, literature performs its own murder. It lives only to mourn its absence, its death at the moment it is born.[4]

Its fate, destiny, or destination aside, literature must, he argued, transcend the matrix of the French Revolution in which it was born. A case must always be made *for* literature: institutions must give the right of way to literature in order to let writers engage in the uneasy, exhilarating, fearful, terrible but powerful doubt that surrounds a craft having neither public end nor constructive purpose. So *Du droit à la littérature*, a title that followed the historical shift "from law to literature," contained in homonymy the plea "for the right of literature." Here, more succinctly and decisively than ever in his career, Derrida stepped forward to defend the cause of an archaic tradition that he had displaced into the field of philosophy.

The lecture averred to be a poetics of sorts—a poetics because, in nearly every literary movement we have known, poetics or *poétiques* have schematized movements already dead and gone.[5] Derrida's plea might well have been related to his own turning to a more literary mode. Who were the major influences and precursors? Above all, there was Maurice Blanchot, who had loosened the Hegelian knot that, in Derrida's view, Lacan had tied around Freud's throat. Then there was Heidegger. The tortuous flow of Heidegger's interrogations, in which the signs of words pass well beyond the grammar containing them, might well have suggested to Derrida the possibility of an endless passage from philosophy to literature and back again. And finally there was Freud himself, especially the Freud who appeared to exceed his usual theoretical bounds in *Beyond the Pleasure Principle*, with its appeal to what the poets have always known. These authors may have prompted Derrida to look to poetry and the plastic arts in order to recast the temper and tone of his own styles. For in Freud, Heidegger, and Blanchot, Derrida found *writers* whose manner determined much of the logic of their matter.

If bibliography can be of help in determining where Derrida's literary writing began, we can note that in 1978 he was composing a second volley of essays on each of these masters. The seminar on Freud would appear in homage to Nicolas Abraham in *Études freudiennes* before being rewritten in

La Carte postale, while his essay on Heidegger and metaphor ("Le Retrait de la metaphore," in *Po&'sie*) can be seen as a supplement to "Violence and Metaphysics" and "White Mythology." And the preoccupation with Blanchot would produce myriad texts around and about *La Folie du jour*.[6]

About the same time, Derrida had published some marginal pieces outside the familiar context of helping students prepare for the *agrégation*. Adrift from academic moorings, or in transit between one institution and another—Yale and the École Normale Supérieure—he wrote for *Digraphe*, a review of fiction and theory, several pieces grafting strips of Blanchot to Ponge, to Freud, and to Heidegger. The latter's sense of irresolute meander in *Holzwege* Derrida could relate to the turn and detour of *Weg* and *Umweg* in Freud's travels to and fro, back and forth, beyond the Pleasure Principle. *Au delà du principe du plaisir* was graphically laced to Blanchot's *Le Pas au-delà* (The Step Not Beyond), which had appeared in 1973; and in turn, the incipit to the novelist's earlier *Celui qui ne m'accompagnait pas*, "Je cherchai, cette fois, à l'aborder" ("I sought, this time, to approach him"), is cut into Ponge's words to heighten the aspect of framing, approaching, boarding, or holding the very objects from which poetry takes as its task to insulate itself. The problem of things cut away or half-detached from the world of objects to be cast into words, what the poet entitled *Le Parti pris des choses*, allowed Derrida to think afresh the questions of words, mimesis, and fetish in fiction, and to play them back through the philosophical texts of Kant and Freud.

Now a strict historical view of Derrida's attraction to literature might regress further, to 1974, with the publication of *Glas*. This double-columned work placed next to each other studies of Hegel and Genet on morality. More astonishing than the identity of these two figures, whom we would never think to include in the same chapter of ethics, much less calculate as reversed surfaces of each other's imaginary topography of behavior, is the fact that Derrida's long academic-philosophic study of Hegel adheres to a supremely poetic one on Genet, but in a particularly graphic way, in which the erotic shape of a double chiasmus or Moebius-strip-twice-over results from the two borders bending back and reading through—doubly invaginating—each other.[7] But even this is not new. Derrida had experimented with sentences reconnecting and perpetuating each other through gutters and paginal borders in his "Tympan" introducing *Marges de la philosophie* (1972), by which a text on the sound and tone of philosophy was stuck next to a strip by Michel Leiris on Persephone. This strategy of collage or montage, of Freudian *Zusammensetzungen* or hieroglyph, had a precedent in the opening to "La double séance" (first printed in *Tel Quel* in 1970, and then placed in the center of *La dissémination* in 1972), where a section of dialogue from Plato's *Philebus* dominates three quarters of a page to which is glued, in the lower right-hand corner, a fragment of "Mimique" by Mallarmé. On first glance the collage turns highly vocal pieces into the silence of emblems; on second viewing the different point sizes and styles of type blend varied *spacings* in the prose in order to body forth a broad range of tone and tension from a single frame of reference—the whole being laid out as if a page could engender multiple voices, hesitations, and redundancies through the mix of time, history, genres, and disciplines that otherwise would be associated with contents apart from the physical gloss of their surfaces.

All of a sudden what has appeared to be an axial moment in Derrida's career looks as if it had been staged all along the way. This is not to imply that his self-conscious writing—and all great styles are self-conscious— had prepared its fields of difference when Derrida was launching a *pro-gramme* (or pro-gramma, an argument for writing) of *lettered* philosophy in the earlier critiques of logocentrism. No, the tension of the *écrivain* was always there; it was only necessary to slip doubt from the protective sheath of the discipline of philosophy or wrench it free of the proleptic manner of Althusserian procedure that had marked so much of the *Grammatology.*[8] Derrida's inclusion of literary styles within reasoned inquiry jolted the monotonous power of logic. It appears as if his task were to displace the mechanism that made of reason a very smoothly tuned machine by momentarily throwing it out of synch with inscription of different melodic inflections. At first Derrida accomplished this by turning philosophical issues against their own mode of formulation. The style of philosophy was not to be understood as a purely literary style, in which the individual beauty of the writer's discourses could account for or valorize itself in some miraculous or ineffable way. Style was to be understood as inscription, as marking in a literal sense rather than as a metaphor of decor. With this "supplemental" view of the craft, "styles" of philosophy and literature would be of the same stamp. In considering the pantheon of Western philosophers, Derrida then juxtaposed the eminence of their reason with the ruses of his best stylists, and in doing so he became what he already had been, without there being any point in his career that we would be tempted to call an anticipation, a shift, a decision, or a phase of development. That is, he became, as he had been foremost, a writer.

In *Spurs*, a lecture on Nietzsche first delivered at Cerisy in 1972, Derrida dropped, as he is wont to do, an ostensive *non sequitur* into the midst of the flow of his reasoning. A clog sabotages the study: *Il faut écrire dans l'écart entre plusieurs styles.*[9] The note summarizes the preoccupation both Heidegger and Nietzsche shared about writing; it also refers to the fact that we all must multiply the *frames* of style from different periods and expand the referential limits we usually impose upon them. There is something here of Proust's advocacy of pastiche as the model for the apprentice to follow in learning how to write. The imitator must find other rhythms through his own combinations of words. He imitates a paradigm or patron (or even pattern) that regresses, at least insofar as the writer imitates a style that had in *its* turn been formed by imitating a pre-text, and so on, all leading to a supreme zero-degree of style, where nothing can any longer be called original. A unique style emerges from its dissolution in and of other models. Any notion of stylistic distortion, deviation, or *écart* from a master that assures a proper name to be associated with a definite turn of phrase (for example, the presence of Balzac's *Illusions perdues* and *La Recherche de l'absolu* in Proust's *A la recherche du temps perdu*) is a paradox. The writer must nonetheless work within it. In weaving different styles and different selves in the same lines, the craft and cadres, the signatures of the stylist have to be lost in the shadings defined by a timeless accumulation of frames and given paradigms. Like metaphor, style must be a cliché of sorts, but its torsions have to be manifold enough to lend its rhythm a broad range of colors, tones, and appeal. Style begins where the writer sees words as a montage of inscriptions and voices, of visible entities and musical spacings.[10]

Derrida visibly writes *il faut écrire dans l'écart entre plusieurs styles* in order, it appears, to recall how the physical act of writing always retraces its own imaginary absence along the cutting edge of an infinitely historical line. Any furrow of words we mark and follow is always written both in serial repetition of itself *and* with an existential urgency. The two opposite conditions are doubly bound to each other. When one of Derrida's voices advocates composition in deviation or derivation (*à l'écart*) from polished style, the gap between himself and—for instance—a Ciceronian model returns to the center of his work at the very moment his frames of cultural reference are pluralized. One must write *in* the margin to reflect the historical perspectives of discourse; but, also, within the remainders—the discards, the leftovers in a game of cards or chance (another sense of *l'écart*)—that determine one's own given (symbolical, political, and historical) conditions as a writer. The signature of the self is erased when it begins to write, so that notions sacred to the order of style find themselves eradicated in the act of writing that reconstitutes them. The individual self enters into a fray of common discourse; it loses the uniqueness of its *écart*, its proper name and personal history, when it represents words in the place of itself. Notions of anticipation, experience, empirical truth, finality of performance, perception, and even the integrity of events are all thrown into question when writing cuts between its own needs and tensions, its urgency and its seriality.

The return of *écart* through *trace* is not a simple palindrome or facile play of verbal shuffle in Derrida's work. If we cast it, like Freud's bobbin, into the rhythms that constitute its discovery, we begin to witness the tonal range of *verbal apprehension* that his writing displays. No sooner do we grasp the linear sense of *trace* than its quadrilateral dimension opens up within itself. The Derridean line no longer has Cartesian invisibility; nor does it conjure up any myth of mimetic volume or spatial presence. It represents whatever its meaning would convey elsewhere, in the domain of repression its movement embodies. Now the seizure of the auto-mimetic effect of *trace-écart* is not without the virtual presence of an un-conscious that permits the speaker (or reader or writer) to use a repressed dimension—*écart* in and of *trace* or *carte*—to lift the limits of language *already* contained in the user's past to project them toward a future con-tained in the genetic dynamics of the here-and-now. Loss of the trace en route to *écart* is manifest evidence of a turn of style; arrival at the frame or deviation brings us to the incorporation of the latent shapes of a Freudian trace. Apprehension of other or different languages in one set of vocables and graphemes allows us to move from a passive relation with language—it usually speaks us—to the active condition of a maker. This involves painful travel to and from all kinds of repression, the uneasiness of loosen-ing moorings and encountering vertigo and nausea, but it has the stake of bringing a reader into a world with new symbolic reflexivity. Derrida coined this in the fragment *Je m'ec* (taken from Genet in *Glas*): when we discern the movement of letters in words that puts into question their formerly unilateral meaning, we also discover the limits of our monocular perspective and the bind of a logic based on the hierarchical rapport of subject to predicate. We take any statement formulated with "There is," *il y a*, or *es gibt*, with caution, for only that fragment or incipit of a sentence is its truth. *Il y a* is akin to *ça me*, which dictates our passive relation with the world. Derrida's citation of *je m'ec* seems to be compounded with Nicolas Abraham's analysis of individuation that takes as its point of departure a passivity called *ça me*, *on me*, or *il y a*, which is turned into *je me*.[11] He uses the frame, the corner of *ec-*, less to constrict than to prod an active role of a user of language with the things that otherwise would define the individual in a matrix of passivity. The user of language writes between *ça me* and *je me*. He or she uses the two for the end of writing with broad tonal range. *Je m'ec* opens onto *je make*, reversing the formerly un-imaginative relation the speaker has had with language. New obstacles, new borders—new *cadres*, new *écarts*—are encountered, as are new repres-sions. But a new, dynamic framing of repressions and progressions allows the maker, the affective subject, to pass through multiple borders that define all lived experience and, indeed, to cope with the world differently. Such a division, effected within the trace of style determined by visible

modes of centering, forces our words to turn into self-generating fragments that divide and reconstitute themselves at the level of single vocables and graphemes.

Freud had called such a montage of figures *Zusammensetzungen*, by
which the sounds that could have been transcribed from a set of minimal
units into unalterable writing find themselves invested with *other* sounds
and meanings. They pass through shapes that originate among combinations of voice, sight, and tracing. Lines of binary thinking that had been
associated with writing (sound and script, primary and secondary process,
conscious and unconscious, thesis and antithesis, sender and receiver, the
logic of day and dream of night . . .) find themselves loosened by the
meanders of style. Shiftings in single words and what free attention[12]
makes of them bring our eyes to an infinitely broader horizon of discourse
not entirely controlled by grammar. Thought jumps along looser knots of
syntax that resist being unravelled into final sets of meaning. Earlier,
Derrida had called this play of order and disorder *dissémination*. Since
then, in mocking the authority this term has acquired in becoming a
meaning taken to be in a dialectical relation with polysemy,[13] Derrida has
invented a term with a more writerly tone that evokes the more varied
paths of the styles of psychic transfer: *dichemination* (*Carte postale*, p. 154).
No longer is limitless dissemination paired with the closed universe of
polysemy; *dichemination* recalls the daimonic urge of writing in its relentless will to divide and detach itself from its sender and receiver. The
neologism depends on the echo of *dissémination*, just as Derrida's term
tranche-fer defines the limits of "transfer," from which it draws inspiration
and on which it has a parasitical dependence. (In *tranche-fer*, the pun on
the axe-blow cuts the very division of dialectic that adepts of psychoanalysis would like to see contained in transference.)

This is why *écart* has been subjected to deconstruction under the pressure of Derrida's writing of the last decade, most notably and decisively in
the pages written on Valerio Adami's drawings. In large typeface and in
margins jutting right to the corners of an issue of *Derrière le miroir* (No.
214, May 1975, in-folio) in which the prose is printed, the author took the
fragment *tr* and retraced it in movements of translation, travel, traction,
trains, traits, transactions, transfer, traversals, trailings, and of course
ritratti. He divested the two letters of their identity unto themselves (*tr*
"heralding" transfer) or their polysemy (the limited number of words that
could be listed under *tr*). Neither an organizing matrix for a closed system
of anagrams about mimesis (*tr*anslation) or drawing (Adami's ri*tr*atto), nor
a unifying grapheme or enunciation—for there is no way to vocalize *tr*—
he let the hieroglyphic fragment, written in the *écart entre plusieurs styles*,
assume a different sufficiency in the dialogue he maintained with his

friend, *Adami son ami*. *Tr* had an "uncanny and haughty" independence that endowed the writer's and the artist's styles with endless movement among ineffable traits, words, or broken lines and figures. Words encountered along the trajectory of *tr* could not be retrieved or recuperated in any thematic inventory, nor could they be reduced to artifacts in Adami's pictures. The letters, words, objects, citations, and backings of the essay reproduce their own play—back and forth. Most powerful is the page entitled "Study for a Drawing according to *Glas*," reprinted in Geoffrey Hartman's *Saving the Text*, in which Derrida's script on the right-hand side is bound to Adami's on the left, within a central border that reproduces a spiral binding of a notebook, between which is a fish hanging from a hook on the upper edge. The presence of *tr* is felt in the bind between the two discourses of the artist and writer.

Tr jumps out of *trace* and spots all kinds of scansions in Derrida's style. It has all to do with the shift from an initial sound that is trayed, translated, and marked over what before had been the indeterminate title of " + *r*". The gap between the crossing X or the *tr* avers to be movement within the difference that Derrida had scripted in the economical title, " + r par dessus le marché," by which the + , a reticle used to border the frame of a page in photo-offset composition, is set in play with the t of *tr*. So: + , X, t, tr, T, x, all keys, double crossings, fabulous markings, and marketings, find themselves theorized and transliterated without the author's recourse to grammar. Offhand incisions, these are what make for the archaic signature, the double self or two writer-artists, in an *écart* of style. The absolute literality of Derrida's writing is most salient when we see it here, in the artist's studio. In the frame of pictures we begin to comprehend how corporal is the notion of *trace*.

We can follow the logic of visibility through two motifs that dominate the writing of the last three years. One involves the art of composing words seen in several languages at once; and the other, the preoccupation with a thematic obsession that could be called—for lack of better words—the knife-in-the-back. The style is haunted by its own penchant to stylize, to produce effects that would make it either imitable or an institutional document of literature. Its earliest and most resonant instance occurs in the arguments of "Parergon," where, in following Kant's third *Critique*, he determines how a third term between structure and decor, a middle or a *Mittelglied* (in which we hear a *middle glide*), puts into question the opposition that formulates it. Lucas Cranach's *Lucretia* of 1533 falls into the text. Is the diaphanous veil over her pubis the *parergon* to the *ergon*, or is it the dagger that "is not part of her natural and naked body whose point she holds back against her flesh, at the contact of her skin (the point of the parergon, alone, then would touch her body at the middle of a triangle formed by her two breasts and her navel"?[14] The point of the style is in the center of an historical triangle, where the triad of our desire—nipples and navel—are matched by that of the knife. The beauty of death, of Lucretia at the verge of her suicide, turning the attribute of Justice back against herself, happens to be a penetration deferred by its image holding it at the point of death.

The same motif comes back through the postcard of Plato and Socrates in Matthew of Paris's fortune-telling book that models for *La Carte postale*. Plato's finger touches Socrates' back, and Socrates reproduces the sign of the pointed index-finger in putting pen to paper with his left hand. An immense phallus is thrust downward, through the back of Socrates' chair and into the buttocks of the philosopher-turned-writer. *Dos*, dos, deux: the two are in complicity since they confront only each other's back. The stiletto would become a sign of the double bind by which the seeming contraries of the philosopher and writer play off and wind through each other in a rapport of death conveyed by means of the process that prints photographic copies of postcards. In writing, Derrida transliterates this as *espee*, the medieval ring of *épee*, a sword with two edges of S and P, of the echo of subject and predicate, of Socrates and Plato, of the writing of a letter that usually defers its most urgent statement to the *p.s.*, of primary and secondary process, and so forth, that are caught in the double dealings of language with its users. In brief, an immense dossier of knives, swords, daggers, rapiers, spurs, and blades could be shaken from this text as analogues to the effects of "style" that run through all of Derrida's writing.

If we were to examine the works closely, as security guards would frisk a known terrorist before an intercontinental flight, other caches of instruments and protective implements of every fabrication would be revealed. Already the critique of simple fetishism that Derrida undertook in the name of *l'argument de la gaine* (the instrument or sheath of gain) in *Glas* (in which he notes that the regressive trait is more economical in its dynamism than as a single fixation upon an object since regression leads us back to the house in which economy, the law-of-the-house, is already established), the metaphor that conveys the problem involves the sheath holding the knife of style which, in turn, must be *self-protective* in order to insure its dissimulations and distortions of words.

Tulips have the same look of terror. In "Parergon," Derrida had chosen to examine Kant's obsession with the sublime through the figure of an artificial flower, a styleless style, a bizarre *sans de la coupure pure*. Tulips do not qualify as really "natural" flowers but as fabrications of the eighteenth century and products of the Netherlands (on postcards). Yet they must find a "natural" frame of reference for their beauty to be imbued with the very culture of an ineffable, sublime appeal. Insofar as the tulip, the most phallic of flora, represents a metaphor of the female—Borgès and others having affirmed this many times—then the vaginal receptable, the *gaine* or sheath does (gains) more than fixing upon or containing its own difference. The tulip is what Derrida likens not just to a matrix in the ideology of puncheon and mold common to naturalism or the dialectical imagination, but to the virtual *possibility* of such a distinction to be conceptualized in the metaphysical tradition. With wry wit he recoins the woman-tulip metaphor as a *conceptacle*, as a potential condition of being-conceived-as-a-concept-of-reception. As such, the new tulip must wilt; it loses its phallic and matrical petals and allows us to think our figure from more than two "points of view." Or, better, it stiffens again after having gone through a weird metamorphosis; in the greater repertory of pointed objects in his work, Derrida's tulip begins to resemble a foil plunged into the earth, the stem a blade crowning a protective guard that might cover a fencer's wrist. The tulip bears the trace of the stiletto in its utterly different conception.

Plusieurs styles, plusieurs langues: Derrida ranks among very few modern writers who attain polyphony through the inscription of many traditions and voices in single words and letters. We think of Joyce as an immediate avatar when we read *dichemination* or *perverformative*, but the novelist's synthetic grids that control works like *Ulysses* do not allow for the drift or travel of words through modern history and human sciences. Or we think of Beckett, whose novels depend on a double scansion of French and English. Or Freud, who reputedly, because of his training in the tradition of the Talmud, could think in four languages at once. It is the marvelous *virtue* of the many registers of Derrida's words that make them defy the schematic reduction to which his commentaries on philosophy have too often been reduced. When his styles are followed carefully, there is no way that the thought can be appropriated in a single tongue.

Like a flower, a hundred words can bloom. Throughout Derrida's recent texts, arguments are begun, sometimes followed to a logical conclusion, and often left aside. "I leave that aside," he says, recalling how the verb *laisser* is accorded a multilingual density. *Je laisse*, he reiterates often through "Restitutions," an essay defending the cause of Heidegger before an imaginary inquisition of Van Gogh by the art historian Meyer Schapiro. The topic at hand was Schapiro's note that in "The Origin of the Work of Art" Heidegger had confused a still-life of a pair of woman's clogs for another of two leather shoes worn by nineteenth-century factory workers. Derrida aims not to "restitute" the philosopher before the seemingly "higher truth" of the art historian but to essay the same *économie de la gaine* of the tulip and instruments of style through the fetishizing of words. Rather than arguing, he constructs a dialogue of several voices that approach, blend, harmonize for a moment, and take leave of each other. The

common term is *laisse:* the laces that distinguish the two paintings, the
rural and industrial economies, and the two shoes become the leash and
laisse by which Schapiro fetishizes the fetishism of Van Gogh's still-lifes—
of objects—detached from the body held outside of the canvas, *à l'écart.*
Derrida leaves, *je laisse,* he says, these remarks aside to respect the signature of the painter. Much of the essay is built around the ways that, in
language, we appropriate things wherever we place—we *laisse* or *lace*—
them aside, like shoes or photographic memories. The utterances we use
to distinguish ourselves from things or situations always wind up reattach-
ing us to them. Now the *laisse* also permits Derrida to place the thread of
Freud's bobbin in *Beyond the Pleasure Principle* back around a context of the
fetish, for in its going-and-coming, in its way that the little boy leaves it to
retrieve it, he happens upon the rhythms of style that are the possible
legacy, or *legs* (which can be pronounced the same as *laisse*), of Freud's
writing. "Legs de Freud," as he entitles the text in homage to the late
Nicolas Abraham (in *Études freudiennes* 13–14), generates *laisse*—or a
leash—of Freud, plus legs (the beat, the measure, the cadence) or legacy of
writing "deferred" ("de Freud," in an English inflection): *laisse* deferred,
legs de Freud, which turn about so that no stable meaning can be derived
from any single viewing or pronunciation of the title.

To repeat, the artistry of style has much to do with these meanders that
can be traced either through translation, a repertory of thematic objects,
or the echoes of a single term such as *laisse.* The knot of *tr* we deduced
from Derrida's inversions of *trace* and *écart* can be used to unfold another
theme or disposition that is refracted through philosophy and literature.
Like *tr,* it is the note of *ec.*[15] EC has no self-sufficient identity; it is the
lapsus, however, among the multiplicities of style that have marked Derrida's most imaginative and most literal pieces of writing. EC, at the initial
edge of a deviation, an ECart, already frames the word before the sum of
ciphers can displace the voice pronouncing *ec: eck,* ec-art, we hear, the
frame or corner, the angle that heralds a frame, or a parergon representing, in delineating what it encloses, the possibility of an imitation, a stasis
or a dose of death. This was the right-hand edge of *Glas,* where *je m'ec*
became the basis of the dialogue of Hegel and Genet over a Rembrandt
torn to shreds and flushed down a toilet. I make, I fabricate, I find the
cornerpiece, the angular limit, the *cadre* or frame of style that would be a
fetish of art and philosophy. The frame of reference has its keystone in the
"bizarre and haughty" independence of *ec* that is seen and heard before—
and simultaneously with—its apprehension in some other verbal unity.

If *ec* is the verbal embossing-punch of *écart,* it also reflects the uneasy,
almost nauseating effect of referential presence and self-identity. For
whatever or whoever can claim to possess something EC—*ici*—here and

now, *hic et nunc*—engages in a relation of power that wills to frame the interlocutor. Whoever pretends to be *ici* envelops others in his own self-presence. An authority that authorizes itself in the name of the place it ascribes to be its identity—as in the indestructible moment that every iteration opens for us through its writing or the confidence we feel when we sign a document with our signature—makes use of surroundings so basic to all mimesis. The power inherent in the use of language is such that every user marks off a space that he takes to be his own; the act of enunciation performs this task of appropriation in his place as he enunciates. In an essay on the way power is self-embodied in language in the models theoreticians of speech use to formulate speech acts, "Signature événement contexte," first published in *Marges* (and then translated in *Glyph* 1, 1978), Derrida had outlined the politics of illocution and defused it through his own graphic sleight-of-hand. The spacing of the title—the two white gaps that lend a coherence to *événement* in contiguity to *signature* and *contexte*—can be turned back to mark tonal stops, virtual commas on the one hand, and, on the other, syncopations that turn the first syllable of *contexte* into a relative pronoun and the second, *texte*, into an archaic verb meaning "to weave" or "to lace": *signature, événement qu'on texte*, or "the signature, an event that we fabricate." With one reading arched over another, the notion of authenticity or singularity (a signature in a given style), the existence of events or empirical truth as a valid science of language, or a context as something apart from what one invests into it for ends of self-interest, is turned into a majestic fiction. When Derrida "agglutinated" his reference to that article in a debate with a group of speech-act adepts at Berkeley (whom he called Searle, Dreyfus, and company), Derrida referred to the origin of the dialogue in what he "dryly" (*sèchement*) abbreviated as SEC. The tempo of the three letters averred to be truer than the staging of the confrontation in northern California, for, in containing the particle EC, SEC became the issue of a failed presence, a non-confrontation in the echo of its literal transcription in *est-ce ici?* Was it, or is *it*, the final iteration about speech-acts, really here? Can dialogue about metaphor and voice ever transcend writing that sabotages them? Where was, where is *it* of the guaranteed referential presence of any speech-act? The style of the ploy was enough to subvert the intended confrontation of two traditions.

Yet it might be argued that Derrida is the architect of the rhetorical scenes and communicational constructs he so readily dismisses; that he models his writing in strict accord with the mode of delivery called for by the contractual conditions of the form, whether an interview, lecture, or any institutionalized situation; that, like so many prominent intellectuals, he puts together books by reprinting, in slapdash fashion, a mass of arti-

cles that cohere only by virtue of an introduction or other prefatory strategy; that the contexts of his work would seem all too predictable in their
blend of philosophy and literature to allow for any reading outside of these
two disciplines. Yet Derrida always uses a given context to affront its
influence and to write against its limits through the temper of style. We
can recall how, in the many interviews since *Positions*, he notes that, here
and there, he must be overly brief and arid, while on some occasions he
must accord himself time and space enough to unwind a thread of reasoning; or that elsewhere he must be brutal and get right "to the point," so
that the tenor of the movement will give definition to the topic in terms of
its pace. Much has to do with recurrent issues of tone, temper, and
rhythm that embody the very concepts of which he speaks. Context becomes a structuring agent for the musical score of writing.[16]

To be sure, the principle of disavowal is crucial to his undoing of the
truth we invest in exchange, intersubjectivity, experience, presence, or
any type of possessive individualism. Any assertion can be turned to speak
for its reverse within it, just as, correlatively, every formulation of law
institutes its desire for transgression of the limits of behavior it describes.
But it is in the area of shaded cancellation of contraries that tonal difference, rhythm, and tension acquire resonance in Derrida's recent pieces.
Here the dialectics are diastolic: they close and open onto possibilities of
style imbued with visible echo of musical writing. Derrida was explicit
about the tone and cadence of his fictional pieces in an important paren-
thesis he opened in a text on Kant that he read at Cerisy two years ago,
and it is here that we can take leave of his preoccupations with style.
Tonion, he reminds us,

> c'est le ligament en tant que bande et bandage chirurgical. La même tension
> traverse en somme la différence tonique (celle qui sous le mot de stricture
> forme à la fois le thème et l'instrument ou la corde de *Glas*) et la différence
> tonale, l'écart, les changements ou la mutation des tons (*Le Wechsel der Töne*
> hölderlinien) qui constitue un des motifs les plus obsédants de *La Carte
> postale*.

> is the ligament as a band and surgical bandage. The same tension in sum
> cuts through tonic difference (that which, under the name of stricture,
> forms at once the theme and the instrument or cord of *Glas*) and tonal
> difference, the deviation, the changes or the mutation of tones (Hölderlin's
> *Change of Tones*) that is the basis of the most obsessive motifs of *The Post
> Card*.

Tone is the last and most indicible element of a unified style. But to
elevate a tone of voice or writing is tantamount to muting other, erotic,
interior voices, the murmurs of others in ourselves, to obliterating the

balances, syncopes, limpid measures of our cadences which make for the fortunes of writing. It appears that much of the recent material has led Derrida to listen to the music of his styles and to measure their rhythms through telecommunications, through novels (of Blanchot and Roger Laporte), and, most of all, with the intransitive practice and daily urgency of his own discipline as a reader and writer. Less and less do we see obvious paragons informing his writing practice.

Yet any introduction to Derrida as stylist cannot avoid comparison. Nor, in his own way, perhaps, can he. Not long ago, Derrida wrote a text in fragments in homage to the late Roland Barthes.[17] Fragmentary as his best writing is—the piece on Adami, certain movements in *La Carte postale*, the Genet column of *Glas*, "Pas," some movements in "Restitutions," and the essays on Freud which are all inconclusive, muting the tones of their assertions or erasing the paths of their inquiries—he uses Barthes's distinction of *studium* and *punctum* that was elaborated in *La Chambre claire* to test the breadth of analogy of the initials *s-p*—the stylus of difference itself that had run across *La Carte postale*. He meant to see exactly how sharp, how acuitous, is the point of *punctum*, but somewhere along the way he drifts from explication of Barthes's opposition to a more fugal play of tones that are at once reverent, critical, melancholy but, above all, more specific and graphic in their meander than what we remember to be the terse elegance of Barthes. It could be that Derrida was simply finding where tonalities could lead him next in the legacy of Barthes, a classic French writer whose short phrases were modelled on the writing of Valéry and Gide. A monument to the urbane and gnomic modes reaching back to seventeenth-century Jansenism, Barthes's style becomes a backdrop for Derrida's bizarre, mannered, tortuous mode: anchored in its own doubt, erasive, evasive, elegiac, jagged, its styles have no roots in French soil. Its pathos seems Germanic, its tragic ironies almost Greek, its jocular play almost cinematographic in an American manner, its graphic immediacy archaic and alphabetical as the Talmud. Only the shell of its logic of the binding of the books recalls the classic look of French masters. Where Barthes's will is to teach through a lucid progression from subject to predicate in which style is its own sensual repression, and the signifier the topic of its signified, Derrida's words ramble and confuse us, they scintillate meaning and trace the erotic edges of their letters. They cut, but unlike Barthes's, nowhere do they lead us to any final sense of their form. All of Barthes's writings calmly teach us the unteachable and shine as pedagogical virtue. Neutral, they do not aspire to the dynamic ranges of Derrida's chosen words and modulated phrases. In a final memo to a comparison of the two figures, we might speculate that Barthes has crowned and celebrated a firm French tradition of restrained style; Derrida has displaced the classical ethos and given to French tensions it could never have felt before.

NOTES

1. Jacques Derrida, "Différance," in *Speech and Phenomena*, trans. David B. Allison (Evanston: Northwestern University Press, 1973), p. 131. Here and elsewhere in the text I have used my own translation.

2. Jacques Derrida, "White Mythology: Metaphor in the Text of Philosophy," trans. F. C. T. Moore, *New Literary History* 6 (1974):6.

3. Jacques Derrida, *La Carte postale de Socrate à Freud et au-delà* (Paris: Aubier-Flammarion, 1980), pp. 181–82.

4. Some of the content of these lectures is derived from Maurice Blanchot's meditations on writing and terror in *La Part du feu* (1949). The context they establish is one of the Age of Terror after the French Revolution. The reason of Terror—its folly has a rigorous logic in its workings—finds its origins in Rousseau's reflection on social problems in eighteenth-century France. These are examined in François Furet's preface to *Penser la Revolution française* (Paris: Gallimard, 1978) and verify the precision of the historical perspective that Derrida elaborates in *Du droit à la littérature*.

5. The history of *poétiques* is reviewed in Albert Thibaudet's *La Physiologie de la critique* (1930; rpt. Paris: Nizet, 1962). Let us recall that the most decisive poetics are those that become their own genre. They announce activities that will never take place, and they schematize movements that will never come true. One of the best examples of a poetics without a practice is Du Bellay's *Deffence et illustration de la langue françoyse* (1549), a manifesto arguing for French by plagiarizing Cicero, Horace, and Sperone in the name of signatory authenticity.

6. Two translated essays by Derrida on *La Folie du jour* are "The Law of Genre," in *Glyph* 7 (1979):176–323; and "Living on: Border Lines," in *Deconstruction and Criticism*, ed. Harold Bloom et al. (New York: Seabury, 1979). He touches on the same text in "Title (to be specified)," *Sub Stance*, 31 (1978):5–22.

7. Schematically, we have:

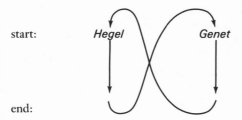

start: *Hegel*   *Genet*

end:

Dichotomies are complicated by the paginal layout. The graphics of this book—a colossus of two colossal stylists—tell us how Hegel closes off his dialectic through the pace and rhythm of his words that advance the meaning they have already played back through writing.

8. It suffices only to reread *De la grammatologie* today. It rings with the rhetoric

of May 1968. The style tends to inflate its hypotheses to the point of ironic hyperbole while, at the same time, it subjects them to historical structures. Like its English translation, the book has a dominantly grandiloquent tone, as when echoing Althusser's theme of an "epistemological rift." Thus, Ezra Pound and Mallarmé were "the *first rupture* of the profound occidental tradition" of logocentrism. Other visionary statements have a Ciceronian air, such as "Ne faut-il pas cesser de considérer l'écriture comme l'éclipse qui vient surprendre et offusquer la gloire du verbe?" ("Must we not keep ourselves from considering writing as the eclipse which happens to surprise and obfuscate the glory of the Verb?" p. 139). Derrida has since written differently. His "Sur un ton naguere apocalyptique" in the colloquium at Cerisy, *Les Fins de l'homme* (Paris: Galilee, 1981), already accounts for this type of style.

9. Jacques Derrida, *Spurs: Nietzsche's Styles*, trans. Barbara Harlow (Chicago: University of Chicago Press, 1979), p. 138. This is a French-English bilingual edition. The full sentence is: "Pour que le simulacre advienne, il faut écrire dans l'écart entre plusieurs styles." Barbara Harlow translates this as: "If the simulacrum is ever going to occur, its writing must be in the interval between several styles" (*Spurs*, p. 139).

10. Until this point, the argument that the tension of style in literary discourse is analogous to the role of metaphor in philosophy would call for extended comparison of literary stylistics to Derrida's "deconstructive" reading of major European thinkers. For if style is determined by its *écart* from a norm, and if its singularity depends on the presence of *cliché*, we would be close to Michael Riffaterre's notion that literature depends upon its reformulation of ready-made statements (studied in *Essais de stylistique structurale* (Paris: Flammarion, 1971). The difference is, however, that in Riffaterre's program of analysis, literature is treated as an artifact exerting no pressure on the discourse of the stylistician. No account is made either of the consequences of its hierarchical or hieroglyphical relation with critical discourse appended to it. With Riffaterre, we may surmise that the logic of its chance, plus the rhetoric of its "unconscious"—the urgency of the actual frame of inscription and the montage of concrete and vocal shiftings that occur in criticism (irrespective of the system of composition or analytical mode)—are elements that his methodologies are hard put to take into account. This is why the brilliant precision of Riffaterre has little to do wtih the problems of writing that are at the very center of Derrida's concern for style. Paul de Man may have come to a similar conclusion when he notes that the stylistician does indeed evade figural problems of writing—its visible, literal, allegorical act of inscription, which is always central to the trace of style in Derrida. See de Man's "Hypogram and Inscription: Michael Riffaterre's Poetics of Reading," *Diacritics* 11 (1982):35.

11. Nicolas Abraham, *L'Écorce et le noyau* (Paris: Aubier-Flammarion, 1978), p. 104.

12. By "free attention" we do not mean "free association." In the former, we actively follow a logic to the consequences it may have never known. In the latter, we are directed, if not *ideologized*, by what we mistakenly feel to be the triumphant polyphonics of our minds' knowing a reason beyond reason. Louis Marin explores the issue in his "theoretical revery" on Giorgione's *Tempest* in *Les Fins de l'homme*, p. 336.

13. Some alert notes on these moves of reading and writing can be found in six paragraphs of "La Pharmacie de Platon," in *La dissémination* (Paris: Seuil, 1972), pp. 110–13. The limits of polysemy are studied in the same volume, pp. 337–38.

Let us recall that most stylisticians use polysemy as a conceptual perimeter: certain words have plural meanings that are controlled by their degree of *écart* that the style provides as a context. In other words, there may be many signifieds for a given signifier, but they remain finite and subject to scientific control. This would be akin to the view of literature we obtain in the guise of literary criticism. Now Derrida is more archaic. With *dissémination* the syntactical play of spacings does not limit polysemy to given combinations of phonemes. The shufflings of meaning are generated by what we ascertain to be the contour of words surrounded by white, or vice-versa; our contamination of sounds of different languages with given scripts; the multifarious crossings that can be made with words through serial play of graphemes that can follow etymology, change in font, and the like. The so-called "free play" is not controlled by some godlike context the analyst imposes upon it, but from given ideological complicities that are in correspondence between reader and text.

14. Jacques Derrida, *La Vérité en peinture* (Paris: Flammarion, 1978), pp. 67–69.

15. To treat the entire unit, we would have to gloss the *a* at the center of tr*a*ce and éc*a*rt. It can be the *a* of différ*a*nce; the vowel of c*a*r, or annulation, displacement, and vehiculation ("tenors" or auto-mobiles) within which writing is written; or the innocuous *car* of French, the "for" or "therefore" that will assure the cause of causality.

16. The patterns of leaving and inventing a context through abbreviation and amplification mark the exchange, in English, entitled "Choreographies," with Christie V. McDonald in *Diacritics* 12 (1982):66–67.

17. Jacques Derrida, "Les morts de Roland Barthes," *Poétique*, 45 (1981):269–92. The text may have been an *envoi* Derrida had purposely sent to a non-existent destination. Since this envoy may have had no receiver in Roland Barthes, it could be that the tension of the discourse as a piece of literature is heard in the tragic context Derrida created in the correspondence with Barthes's photographic residue. "An *envoi* may not arrive at its destination, but precisely it is there, in this drama, in this tragedy, that are found the origins or the possibility of literature" (*Les Fins de l'homme*, p. 213). The writer is mourning the function of writing-as-melancholy to a second degree. Derrida performs the task of literature by *staging* the function of mourning that is taken to be the modern condition of writing (a theme explored at length in the work of Michel de Certeau, Jean-François Lyotard, and, of course, Barthes himself in his relation to his mother). In "regretting" Barthes, Derrida displaces the dead body into words of his own fabrication. The essay is not unlike the staging of loss (for gain) that Montaigne in his essay "Of Friendship" performed in relation to the death of his friend Etienne de la Boetie.

I wish to express gratitude to the Camargo Foundation (Cassis, France), which provided the time, leisure, and Mediterranean perspective for my writing of this text, and to Verea Conley for her critical reading of it.

Part Two

***

Negative Theology, Heretic Hermeneutics

Introduction:
Negative Theology, Heretic Hermeneutics

"JACQUES DERRIDA AND HERETIC HERMENEUTICS," by Susan Handelman, poses peculiar interpretive problems for the reader. One way in which these problems may be broached is through considering the relationship between negative theology and heretic hermeneutics. Negative theology may be defined as that theology which is informed by radically negating moments and events. Heretic hermeneutics, on the other hand, denotes a mode of interpretation which may function either at the margins of tradition or by transgressing these boundaries entirely. By orienting our reading in these terms, we find three critical moments running through Handelman's essay: negative theology verging upon "heresy," heresy becoming "orthodoxy" and thus normative, and the attempt to develop criteria for assessing such shifts between negative theology and heretic hermeneutics.

Handelman's point of departure is the fluidity of boundaries between text and commentary in both the Jewish hermeneutical tradition and in contemporary literary theory. The question raised by her own work is how, in a tradition which is constructed through displacement of the text by commentary, we can generate criteria that will locate the difference between a normative and a heretical hermeneutical displacement. Handelman approaches this problem by characterizing heresy as a "displacement displaced." The heretical impulse undoes the Torah through the Torah: "Rabbinic displacements . . . have been used to overthrow and displace the Rabbis." Heretical hermeneutics, according to Handelman, subverts the dialectic of identification and displacement into a displacement of the displaced. Handelman uses a dual focus to explore this dynamic: (1) she locates the functioning of heretic hermeneutics both within its own tradition of heresy as well as *within* the Rabbinic tradition itself; and (2) she analyzes Derrida's strategy of countering the logocentric tradition of Greek and Christian hermeneutics with the tradition of Écriture.

Handelman's interpretation of Derrida as Reb Derissa treats his writings as a form of heretic hermeneutics within a larger tradition of heresy. She carves out a space for Derrida within what she terms the Rabbinic dimension of the Jewish hermeneutical tradition after that tradition has taken a final heretical turn. Handelman reveals herself to be a strong reader in privileging this heretical tradition. For this move, she invokes the authority of Gershom Scholem, the historian who has transformed our sense of the relations in Judaism between esoteric and exoteric traditions. Scholem would probably call the esoteric tradition "marginal" rather than "heretical." In any case, Handelman, in privileging the heretical tradition, herself enters very closely into just the kind of reversal between heresy and orthodoxy she is ostensibly explicating. Indeed, one might argue that by making such a strong case for the intrinsic character of the heretical tradition as arising out of the Rabbinic tradition, Handelman has from the outset set up the possibility for just this final reversal between heresy and orthodoxy. In Handelman's strong reading, the displacing Rabbinic tradition prefigures the heretical, and the heretical tradition is always already, in some ways, Rabbinic.

Handelman does not here specify what Rabbinic hermeneutics might be. She does take this up in detail in her book, *The Slayers of Moses* (1982), particularly in chapter 3, pp. 51–82. There she notes the fundamental distinction made within Talmudic discourse between two kinds of argument, *Aggadah (Midrash)* and *Halacha*. The activity of infinite displacement more properly belongs within the context of aggadah or the extra-Talmudic writings of midrash, rather than within halachic writings. The similarities and differences between these two modes of inquiry are themselves a topic within Talmudic discourse, and how one distinguishes between these interpretive functions shapes how, in turn, one distinguishes and relates the normative Rabbinic and heretical hermeneutic modes. In this essay, it seems that the more thoroughly displacing aggadic interpretive mode is given centrality in characterizing Rabbinic hermeneutics. This move has significant consequences for the construal of the dialectics of orthodoxy and heresy in the Rabbinic tradition.

Handelman's reversal of heresy and orthodoxy, whereby the marginal becomes central and thereby normative, is accomplished in three steps: (1) the characterization of the Rabbinic tradition as constituted through, and thus *intrinsically*, displacement; (2) the exacerbation and *intensification* of this displacement by considering it the dominant trope of the Rabbinic tradition, displacing all other tropes, and finally displacing the boundaries between heresy and orthodoxy; and (3) the final *reversal* of heresy and orthodoxy, legitimating heresy in terms of displacement as that which is paradigmatic, definitive, and constitutive of the Rabbinic tradition.

This final reversal is implied rather than stated, at the end of Handelman's essay. It is indicated by her use of Scholem, as already noted. But it may well be argued that this final inversion is systematically implied in the basic terms, assumptions, and methodological strategies she uses throughout her argument. It appears that Handelman's own method partakes of the heretical hermeneutics she seeks to explicate. Heretic hermeneutics is not simply her subject matter. For in what we may call (after Harold Bloom) a "strong reading," she both produces and participates in what she is ostensibly only seeking to elucidate.

Handelman's essay, then, makes use of the same strategies of displacement and reversal that she explicates in the writings of Derrida. By subverting and transgressing the boundaries between heretic and normative Rabbinic hermeneutics in terms of a "displacement of the displaced," Handelman, as we have seen, is able not only to consider displacement as irreducible, in the sense of an always and already present dimension of Rabbinic hermeneutics. She may then consider the final displacement that characterizes heretic hermeneutics as a paradigmatic Rabbinic hermeneutic. This locating of the figure and function of displacement within orthodoxy blurs the boundaries between orthodoxy (the Rabbinic tradition) and heresy (Derrida). Handelman makes an intrinsic unity of these supposedly mutually exclusive forms of interpretation.

When writing itself is the subject matter of writing, and when reading is always and already to write, the scene is set in which displacement may or even must occur. It is not surprising, then, that Handelman's writing on displacement calls forth the very dialectics of displacement, between negative theology and heretic hermeneutics and between writing and reading, about which she discourses. Because writing on the subject of displacement partakes of the thoroughgoing displaced character of all writing, even that which is written about displacement is displaced in writing, and must be further displaced in reading.

—SUSAN E. SHAPIRO

JACQUES DERRIDA AND THE HERETIC HERMENEUTIC

***

Susan Handelman

The Jew is an expert at unfulfilled time.
—ARTHUR COHEN

In Algeria, in the middle of a mosque that the colonists had changed into a synagogue, the Torah once out from *derrière les rideaux*, is carried about in the arms of a man or child. . . . Children who have watched the pomp of this celebration, especially those who were able to give a hand, perhaps dream of it long after, of arranging there all the bits of their life.

What am I doing here? Let us say that I work at the origin of literature by miming it. Between the two.[1]

THE SPEAKER IS DERRIDA in an autobiographical passage in *Glas*. And we too may ask, indeed, what *is* Derrida doing here—unveiling the veiled Scroll, the veiled Writing, and here arranging all the bits of his life? Gayatri Spivak interprets this passage as the Jewish child's inspiration at the absence of the Father, or Truth behind the veil, an inspiration that allows him to place his autobiography in that place, producing the "origin" of literature.[2]

To place his own autobiography in place of the Torah is, first of all, to displace that most primary and authoritative Jewish Text—and yet, at the same time, paradoxically to continue it, "to arrange there all the bits of his life." And this Derrida does with his all-embracing theory of *Écriture*. Usually translated as "Writing," *Écriture* can also be translated from the French as "Scripture." Most readers of Derrida interpret his displacements, dislocations, and mimings as extreme extensions of a Nietzschean-Heideggerian tradition of "de-construction." I would maintain, however, that they need to be understood just as much as an extreme extension of a long tradition of Rabbinic Scriptural "heretic hermeneutics" that includes figures as diverse as the apostle Paul, Freud, and Harold Bloom. And this heretic hermeneutics is itself generated by the intrinsic displacements of Rabbinic thought.

"Displacement" may, in fact, be taken as a key term for Jewish hermeneutics in general, and needless to say, constitutes the Jewish historic condition of Exile. Derrida, in his essay on the French-Jewish writer Edmond Jabès, traces this connection between the Jew-as-exile and writing:

> . . . in question is a certain Judaism as the birth and passion of writing. The passion *of* writing, the love and endurance of the letter itself whose subject is not decidably the Jew or the Letter itself. . . .
>
> The exchange between the Jew and writing as a pure and founding exchange, an exchange without prerogatives in which the original appeal is, in another sense of the word, a *convocation*. . . . And through a kind of silent displacement toward the essential. . . , the situation of the Jew becomes exemplary of the situation of the poet, the man of speech and writing. . . .
>
> The Poet and the Jew are not born *here* but *elsewhere*. They wander, separated from their true birth. Autochthons only of speech and writing, of Law. *"Race born of the book"* because sons of the Land to come. . . .
>
> The necessity of commentary, like poetic necessity, is the very form of exiled speech. In the beginning is hermeneutics.[3]

The poet and the Jew are not rooted in any empirical, natural present; they are never *here*, but always *there*, continues Derrida; they wander in the Desert, where the immemorial past is also the future. They are natives only to the word and to Scripture; the home of the Jews is a sacred text in the middle of commentaries. The Jew, says Derrida, chooses Scripture (Writing-*Écriture*), which chooses the Jew. The poet, too, is "chosen," selected by words, and engages in an arduous labor of deliverance by the poem of which he is the father. The poet, too, is the subject of the book, its substance and its master, its servant and its theme; and the book is the subject of the poet.[4]

To be both master and servant of the Book is itself a paradoxical condition; it defines the creative tension of the Rabbinic relation to the text. For the Jews as the "People of the Book," the central issue is how to deal with a canonical, divine Text that claims to be the essence of reality. In other words, the central problem is that of interpretation. (In Harold Bloom's terms, how does the belated commentator-critic-poet deal with the overwhelming influence of a prior, authoritative, Paternal Text?[5]) As a solution, the Rabbis created a system of interpretation that *itself* became another equally authoritative canon, another Scripture. They called it the "Oral Torah" (though it has been written down for about 2,000 years now), and it is their own accompanying interpretation, amplification, and debate over the meaning of the Written Torah—the Old Testament Scriptures. (By "Rabbis," I am here referring to that group of sages, scribes, and interpreters from Ezra after the return from Babylon in the fifth century B.C.E. to the completion of the Babylonian Talmud, the "Oral Torah"—in approximately the fifth century C.E.)

THE TWO HOUSES OF ISRAEL

In Simon Rawidowicz's words, the Rabbis of this period created, in effect, a "Second Beginning," a "Second House" of Israel, in contrast to the "First Beginning" and "First House" of the original Scriptures.[6] Their stunning and audacious achievement was to make this Second House, this Oral Torah (their commentary and interpretation), not only equal to, but in some ways even *more* authoritative than the original Scripture of the First House (a species of displacement again).

There is a famous Talmudic passage which well describes this attitude. R. Eliezer was disputing with the Sages on a question of whether a certain oven was ritually clean or unclean:

> On that day R. Eliezer brought forth every imaginable argument, but they [the other Rabbis] did not accept them. Said he to them: "If the law agrees with me, let this carob-tree prove it!" Thereupon the carob-tree was torn a hundred cubits out of its place—others affirm four hundred cubits. "No proof can be brought from a carob-tree," they retorted. Again he said to them: "If the law agrees with me, let the stream of water prove it!" Whereupon the stream of water flowed backwards. "No proof can be brought from a stream of water," they rejoined. . . . Again he said to them: "If the law agrees with me, let it be proved from Heaven!" Whereupon a Heavenly Voice cried out: "Why do you dispute with R. Eliezer, seeing that in all matters the law agrees with him!" But R. Joshua arose and exclaimed: "It is not in heaven." What did he mean by this? Said R. Jeremiah: That the Torah had already been given at Mt. Sinai; we pay no attention to a Heavenly Voice, because Thou hast long since written in the Torah at Mt. Sinai, *"After the majority must one incline."*
>
> R. Nathan met Elijah and asked him: "What did the Holy One, Blessed Be He, do in that hour?—He laughed, he replied, saying, "My sons have defeated Me, My sons have defeated Me." [*Bava Metzia* 59 a & b]

This story piously and figuratively illustrates the inner dynamic of Rabbinic thought: it contains within itself the possibility for its own interpretive reversals (despite any Divine Voice to the contrary—in fact, poor R. Eliezer, despite the backing of Heaven, was excommunicated after this incident). These reversals are accomplished, above all, by acts of revisionary interpretation (displacements) which take the form of and conceive themselves as another pious commentary on the text. For example, the text of Deuteronomy 21:18–21 specifically declares that a rebellious son must be stoned to death. The Rabbis, however, claimed that the law applied only if the son committed the transgression within three months of the age of 13 and only if the trial were completed in the same time, thus making it practically impossible to apply the punishment. About this verse they wrote: "There never has been a stubborn and rebellious son, and never will be. Why then was the law written? That you may study it and receive reward" (*Sanhedrin* 71a).

There is, however, a subtle inversion of text and commentary occurring here . . . until finally the Rabbis proclaimed: "All that a brilliant student will in the future expound in front of his teacher was already given to Moses at Sinai" (*Yer. Peah* 6:2). That is, all later Rabbinic interpretation shared the same divine origin as the Torah of Moses; interpretation, in Derridean terms, was "always already there." Human interpretation and commentary thus become part of the Divine Revelation! The boundaries between text and commentary are fluid in a way that is difficult to imagine for a sacred text, but this fluidity is a central tenet of contemporary critical theory, especially in Derrida. Geoffrey Hartman, for example, propounds the view that any absolute distinction between literary criticism (commentary) and literature (Text) is naive, that "There is no absolute knowledge but rather a textual infinite, an interminable web of texts or interpretations."[7]

> The line of exegesis will therefore tend to be as precariously extensible as the line of the text. The subject matter of exegesis, is, in fact, this "line." Yet criticism as commentary *de linea* always crosses the line and changes to one *trans lineam*. The commentator's discourse, that is, cannot be neatly or methodically separated from that of the author: the relation is contaminating and chiastic; source text and secondary text, though separable, enter into a mutually supportive, mutually dominating relation.[8]

The question of contamination and transgression is important. When does the contaminating commentary become inversion and displacement? And, returning to our theme of the Rabbinic heretic hermeneutic, when does the radical revisionism of Rabbinic thought become antithetical? Carried to the extreme, or under the pressure of historical or personal catastrophe, and the need to find meanings within Scripture to accord with contradictory contemporary experience, commentary and interpretation can edge over into heresy—under the guise of the extension of the canonical, or even as open rebellion. Christianity, for example, undoes Rabbinic law, all the time claiming to be the true extension and correct interpretation of the now "Old" Testament. Within Judaism, the false messianic claims of Shabbatai Zevi in the seventeenth century similarly asserted a new fulfilling interpretation of the Torah; and there are numerous other examples of schisms, factions, and inversions—some remaining within Judaism, others breaking with it entirely.

The important point is that the factor which accounts for the essence, the vitality, the creativity of Rabbinic Judaism (and which may be the secret of Jewish survival in Exile) is also that which can lead to its undoing: it is "displacement displaced." Every Jewish heretic hermeneutic has undone the Torah through the Torah—be it the interpretations of Paul, or of Derrida. Rabbinic displacements, that is, have been used to overthrow and displace the Rabbis. In sum, the heretic hermeneutic is a complex of identification and displacement; it is inextricably linked to a Jewish Scriptural and exegetic tradition which it inverts, yet which somehow retains a compelling power.

The Rabbis accomplished in their "canonical" hermeneutic, though, in Rawidowicz's words, a "revolution from within," freely re-shaping and re-creating the Scriptures that had been handed down to them. Their mode of interpretation, he asserts, stands for a model of interpretation in general, because their battles, like every interpreter's, were born of the tension between continuity and rebellion, attachment to the text and alienation from it. They teach man how to "uproot and stabilize simultaneously, to reject and preserve in one breath, to break up and build— inside, from within, casting a new layer on a previous layer and welding them into one mold (which later became the great problem of Jewish thought and being)."[9]

The central question here is, however, how does one contain the revolution to the "inside"? What is the line between inside and outside, and what happens when commentary crosses that boundary? This indeed became the great problem of Jewish thought and being—especially for the post-Enlightenment Jews who have, like Freud, like Bloom, like Derrida, assimilated into the outside, into Western secular culture, who have absorbed Western philosophy yet still dream of that ceremony of the Torah, of arranging there all the bits of their lives, who dream of an all-encompassing Scripture, a Writing that weaves together the fragments of reality and simultaneously disseminates endless new meanings through its interpreter sons.

For this reason, Derrida, as part of this heretic hermeneutic, is obsessed (like Freud and Bloom) with the question of origins, and with the need to undo, re-write, or usurp origin—above all, through acts of revisionary interpretation. This is, of course, also a displacement of the "father"—the authoritative, originating principle. Derrida's target is all the fathers of philosophy. His project: to deconstruct the entire Western tradition of "onto-theology," to undo "logocentrism," to send the Word into the exile of writing.

SCRIPTURE VS. LOGOS

It is in the first part of *Of Grammatology* that Derrida most clearly articulates his theory of Writing. In his view, the whole history of metaphysics from the pre-Socratics to Heidegger has assigned truth to the logos and "debased" and "repressed" writing. The spoken word has been considered closer to the immediacy of "inner" truth, and the written word seen as merely second-hand, an exterior and insubstantial double, a "fallen secondarity." Derrida intends to break the link of truth with logos and destroy the science of signification that privileges the phonic signifier, thereby redeeming Writing from its fallen condition. Above all, Writing is the realm not of Presence—to which the voice is so intimately linked—but absence, deferment, and difference, all of which become value-terms for Derrida. Phonocentrism and logocentrism, on the other hand, merge with the determination of being as presence, whether it be the presence of thing to sight, or presence as substance, essence, existence—or the self-presence of the cogito, consciousness, or subjectivity, and so forth.

The immediacy of speech, the proximity of voice and being, lends itself to the privileging of the *phonè* as the nonexterior, noncontingent, nonempirical signifier. A whole series of oppositions thereby arises: the difference between worldly/nonworldly, outside/inside, ideality/nonideality, universal/particular, transcendental/empirical.

Derrida, moreover, at every point indicates the "onto-theological" basis (and bias) of Western thought, the effects of theological thinking on our concepts of meaning, interpretation, and truth. The very act of differentiating between signifier and signified and postulating their exteriority to one another belongs to the "epoch of Christian creationism and infinitism when these appropriate the resources of Greek conceptuality."[10] Derrida is at pains to uncover the "metaphysical-theological roots" of sign theory, a theory entirely dependent in his view on the distinction between sensible and intelligible. This distinction leads to a differentiation of signifier and signified and postulates a signified which is purely intelligible—an absolute, ideal logos, or the infinite creative subject in medieval theology: "The sign and divinity have the same place and time of birth. The age of the sign is essentially theological."[11] In this epoch, the signifier is castigated as a mediated exteriority, which becomes the exteriority of writing in general. The signified is always related immediately to the logos and only mediately with the signifier, i.e., writing.

What Derrida says is doubtless true in the Christian tradition. Rabbinic thought, however, can be characterized precisely by its "escape" from this Greco-Christian onto-theological mode of thinking. For the Rabbis, Writing, the Text, not only precedes speech but precedes the entire natural world. Rabbinic thought does not move from the sensible to the ideal transcendent signified but from the sensible to the Text.

For instance, the distinction is often made between the Hebrew *davhar* and the Greek *logos*, which subsequently became the logos of the famous opening of the Gospel of John, "In the beginning was the logos."[12] *Davhar* means both "word" and "thing," but thing not as *res*, substance, but as "essential reality." For the Hebrew, language is not an exterior realm (as it is for the Greek), an imperfect, imitative realm, but constitutes the essence of reality. Scripture is far more important than Nature. The Greek ontology that Derrida attacks, on the other hand, separates language from being, and steers the truth-seeker towards a silent ontology.

Plato, of course, banished the poets from his Republic and speaks disparagingly of writing, especially in the *Phaedrus*. The progression he requires is from word to thing to form; Being must be known from itself, not from language. The Christian Word as well is, finally, a Word that "fulfills" and transcends language through the central doctrine of the Incarnation. The word becomes flesh, ending the long agon of Rabbinic interpretation and superseding the now "Old" Testament for the "New" fulfilled one. For the Rabbis, however, Interpretation, not Incarnation, is the central divine act.

Rabbinic thought was always an "alternate metaphysics." In part, this was due to the Biblical doctrine that the world was created, not, as in the Greek view, eternally existent. And this view, as Hans Jonas points out, contained an intrinsic "antimetaphysical agent" that "led to the erosion of classical metaphysics, and in its outcome changed the whole character of philosophy . . . the Biblical doctrine pitted contingency against necessity, particularity against universality, will against intellect. It secured a place for the contingent within philosophy, against the latter's original bias."[13] In the Biblical view, because there was no *necessary* existence, everything was contingent.

Rabbinic logic, likewise, could not be founded on necessary axioms inherent in the nature of things that furnished neat syllogisms and universally true statements. Instead, it was dialectical, conditional, wary of universal statements, attentive to the claims of the particular, and relentlessly skeptical—even of reason.

Based on the central principle of Being, however, Greek logic concentrated on the relation between subject and predicate connected by the copula *is*. Rabbinic logic, by contrast, focused on relations of juxtaposition, contiguity, association. Hebrew does not have any form of the verb "to be" in the present tense. Predicative utterances are linguistically constructed through the juxtaposition of nominal forms in a free order—and this linguistic structure may underly the Rabbinic logical principle of predication by juxtaposition.[14]

Moreover, claimed the Rabbis, the world itself was created through the Text: "God looked into the Torah and created the world" (*Bereishit Rabbah* 1:1). The Hebrew *davhar* was an aspect of the divine creative force. Christianity took this non-ontological concept of a pre-existent Torah through which the world was created, combined it with Greek concepts of substance and Being, and developed the incarnate logos. Logos originally meant "to gather, arrange, order" and had nothing to do with the function of speaking; it meant, rather, a rational, ordering principle. In Christianity, this nonlinguistic ordering principle then became a theophany, making a visible appearance in the Incarnation.

And, of course, here is also the standard contrast between the Greek emphasis on seeing and the Hebraic emphasis on hearing. Seeing is presence, fullness; hearing implies absence. When sight is the predominant mode, then in the search for identity in knowledge resemblance will be defined in terms of copy, re-presentation, and thought is *specu*-lative—speculative as specular. And this view of knowledge and mimesis is exactly what Derrida and the deconstructionist critics attack.

THE LETTER AND THE SPIRIT

The Greek view of language as a sign that represents by abstracting from the particular and pointing towards the invisible is part of the whole larger movement from sensible to nonsensible that according to Derrida is one of philosophy's cardinal sins. Following Heidegger, Derrida claims that the metaphysical exists only in the realm of the metaphorical[15]—that equivocal, nonsyllogistic, linguistic realm where a term both "is" and "is not" at the same time. Paul Ricoeur perceives a kind of logical heresy in metaphor, and calls it "aberrant attribution," "categorical transgression," "semantic impertinence."[16]

In the Aristotelian (*Rhetoric* 1457b) view, however, the figural meaning of metaphor was seen as a sort of transfer and deviation from the "proper" meaning, for which it was substituted. The ultimate determinant of what was "proper" was the true predicate of all things, the foundational principle of Greek metaphysics, the central *ousia*, the univocal being beyond language that nevertheless allows the entire system of words and things to operate. The restoration of the proper meaning in effect "cancels" the figural meaning which is merely an ornament, an alien deviation which, by virtue of a certain resemblance with the proper meaning, acts as a substitute for it.

This metaphorical transfer from the "proper" to the "figurative" sense, according to Heidegger and Derrida, is based on a metaphysical transfer from the "sensible" to the "nonsensible" realm, a transfer so determinative for Western thought. The entire ontological tradition of Western metaphysics is based on the Platonic transfer of the soul from the visible to the invisible world. Sensible to nonsensible then becomes literal to figurative, and then, in Christian thought, the letter versus the spirit.

One of the central debates between Judaism and Christianity was the letter versus spirit issue. Christianity castigated Judaism as the religion of

the dead letter, as a perverse and blind devotion which refused the illumi-
nation of the "spiritual meaning" which came through Jesus. Paul puts it
most bluntly: "For the Letter kills, but the Spirit gives life" (2 Cor. 3:6). In
his more generous moments, though, Paul interprets the letter, the law, or
the written code as the temporary "figure" or "type" of the fulfillment to
come in Jesus. The spiritual interpretation thus returns the "proper"
meaning to the text, making all that had come before (the Old Testament
Scripture) a mere shadow compared to the substance (literally) that Jesus
represents. Jesus becomes the true predicate of all statements, the singular
and ultimate referent, a referent entirely beyond language whose appearance nullifies the Jewish scripture, which is now considered as a long
deviation, detour, exile. The Oral Torah was entirely invalidated.

Paul and the Church Fathers after him replaced the prolonged Rabbinic
meditation on and mediation of the Text with the pure unmediated presence of Jesus, who resolves all oppositions, stabilizes meaning, provides
ultimate identity, and collapses differentiation. Language, argument, play
of difference hinder the immediacy of direct union. With the restoration
of the substance, the shadows disappear, and what is left is direct union
with the presence of Jesus. Paul is not only antinomian, he is also anti-
textual. He is impatient and frustrated with the Jews because they cling to
the "letter"; that is, they will not read the text Christocentrically: "But
their minds were hardened; for to this day whenever Moses is read a veil
lies over their minds; but when a man turns to the Lord the veil is removed" (2 Cor. 14–17).

The Jew, that is, will not abandon Scripture for Logos. The Jew approaches divinity through ever more intense concentration on the text—
not through a transcendent vision of the spirit that ends the labors of
interpretation. The text, for the Rabbis, is a continuous generator of
meaning, which arises from the innate logic of the Divine language—the
letter itself. Meaning is not sought in a nonlinguistic realm external to the
text. Language and the text, to use Derridean terminology, are the place
and play of differences; and truth as conceived by the Rabbis was not an
instantaneous unveiling of the One, but a continuing process of interpretation which God himself participates in, learning what the Rabbis have to
say.

The Jews thought of the word-made-flesh doctrine as a pagan literaliza-
tion and blindness. And paganism has precisely to do with ontology.
Christians ontologize the *davhar*, they "literalize the metaphor." They
undo the difference between God and man, Scripture and Nature. Derrida likewise sees the sin of Western philosophy and science as a literaliza-
tion of the metaphor, and a metaphorization of the literal. Or rather, in
undoing *ousia*, Being, the primary predicate and essence of Greek ontol-

ogy, he also undoes the metaphysical opposition between proper and figurative meaning. For both Derrida and the Rabbis, that is, idolatry is the *reification* of signs, image-making, applying ontology to textuality.

Augustine, for example, an exemplary Christian theologian, accuses the Jews of stubbornly adhering to signs—instead of the *things* they are supposed to signify, and refusing to believe in Jesus "because he did not treat the signs in accord with Jewish observance" (*On Christian Doctrine* III:6). Christian liberty, he says, is freedom from signs, "elevating" them to the things they signified. Signs can thus become sacraments, invocations of the presence of that to which they point; they are thus "fulfilled" in the presence of the "thing" which cancels the now "empty word." Jesus is the end of the law.

But for the Jew, precisely this cancellation of the sign for the thing is idolatrous. John Freccero expands on Yehezkel Kaufmann's insight:

> The Jewish concept of idolatry was a kind of fetishism, the worship of reified signs devoid of significance. The gods of the Gentiles were co-extensive with their representations, as though they dwelt not on Olympus or in the skies, but within a golden calf, or stone or piece of wood. Signs point to an absence or a signification yet to come. . . . Idols, as the Jews understood them, like fetishes, were a desperate attempt to render *presence*, a reified sign, one might almost say a metaphor. It is almost as if the Gentiles, in the Jews' reading, sought to evade the temporality inherent in the human condition by reifying their signs and thereby eternalizing significance in the here and now.[17]

Derrideanism, too, is a stubborn adherence to the free-play of the sign, a refusal to stabilize it, or to posit univocal referents that fulfill it. In an Augustinian or Pauline view, such linguistic multiplicity, however, is a condition of the fall. The loss of a stable referent that grounds the "literal" and "proper" meaning of words is a manner of exile; language is a detour, a regrettable mediation, an interference between man and God.[18] Floating signifiers, the play of interpretation, are departures from the "real" referents. Derrida's play of difference, the endless interpretation and commentary of the Rabbis, are for the Christian an unacceptable exile—a displacement from true being. Moreover, in the words of the Church Father Cyril of Alexandria:

> The Jews are the most deranged of all men. They have carried impiety to its limit, and their mania exceeds even that of the Greeks. They read the Scriptures and do not understand what they read. Although they had heavenly light from above, they preferred to walk in darkness. They are like people who had neither their mind nor their thinking faculty. Accordingly they were seized by the darkness and live as in the night.[19]

This was a reading controversy with a vengeance. What collusion with Satan does the Jew have, the Christian wonders, to exist so well within the realm of difference, in the infinite regression of signs, in the cacophony of words and interpretations, in the endless referentiality of the letter, without the redeeming ultimate presence of the Word?

DISRUPTING THE HOLY FAMILY

Interpretation, mediation, displacement, deferment, exile, absence, equivocal meaning—these are the themes of Rabbinic and Derridean interpretation. For the Jews, as for Derrida, there has been no redemption; there is no fulfilling presence. And nonfulfillment, as Jean-François Lyotard points out, is a characteristically Jewish mode. Judaism is not, as he writes, a religion of reconciliation of son and father, where the Other (Father) is returned to the Same. In Judaism, there is between father and son what Lyotard calls an "alliance, a preconciliation" because the son is possessed by the father's voice, through the gift of the text. Christianity is "Oedipal" in the sense that it fulfills the desire of the son to take the place of the father, and Oedipus' desire and his fate coincide. Jesus, in representing the desire to replace the father, bears the guilt of the desire through his death, but simultaneously is transformed into father. In Judaism there is no such conciliatory dialectic.[20]

The God who speaks "dispossesses" (displaces), so to speak, the subject. Just as Oedipus is dispossessed of his origin in Greek tragedy, so too is the ethical—or Jewish—subject dispossessed by being "chosen" through the imperative of the Divine word. The difference between himself and the Other remains; it does not collapse in an erotic drive to bring the Other back to the Same. In Jewish thought, the difference between the father and son is irrevocable. There is no "fulfillment" of the word in an ontological return to the sameness of son and father as in Christianity. The writing, the text as gift, is the father's presence-in-absence. Through the text, the subject is taken and possessed—the son is possessed by the voice of the father.

In Lyotard's view, then, the difference between Greek fate (the tragic) and Jewish kerygma (the ethical) is the difference between Oedipus and Hamlet: the representation of desire fulfilling itself in nonrecognition (Oedipus), and desire not fulfilling itself in compulsive representation (Hamlet), where the subject is displaced in relation to his desire.[21] For Lyotard, the difference between modern and Greek thought is precisely this issue of fulfillment of the paternal word.

It is also interesting to note here that much of Derrida's discussion of Hegel in *Glas* centers around Hegel's early text *The Spirit of Christianity*, where Hegel discusses Christianity as the religion of the Holy Family par excellence. According to Gayatri Spivak's reading of *Glas*, Derrida claims that precisely what the Jew cannot understand is the relation between the Christian and the Divine Father, especially in its combination of the finite and the infinite: " . . . what he does not understand . . . [is] the commensurability of the passage between the two, the presence of the immense in the determinate, the beauty and immanence in the finite."[22] Spivak adds that in the process the Jew is perhaps "denied" any understanding of the philosophical proposition. Says Derrida:

> In every proposition, the binding, agglutinating, ligamenting position of the copula *is* conciliates the subject and predicate, interlaces one around the other to form one sole being. . . . Now this conciliation which supposes— déjà—a reconciliation, which produces in a way the ontological proposition in general, is also the reconciliation of the infinite with itself, of God with Himself, of man with God as a unity of father-to-son.[23]

The Jew is unreconciled: he exists in the realm of desire, difference, and displacement—not fulfillment, identification, and unity. Hence it is no wonder that in his essay on Jabès, Derrida describes the exchange between the Jew and Scripture as a "long metonymy." Metonymy, in the linguistic sense Roman Jakobson has given it, means the contiguous, serial play of signifiers, through combination and contexture. Metaphor is the arrangement of the sign through substitution and selection.[24]

HERETIC HERMENEUTICS: IDENTIFICATION AND DISPLACEMENT

Now we can further understand the displaced Rabbinism of Derrida and the Jewish heretic hermeneutic. We can think of displacement in two ways: as a metonymical repositioning in the manner of contiguity, or as metaphorical mode of cancellation in *substitution* of one term for another (figurative for literal). And in this sense, metonymy can be said to roughly characterize Rabbinic thought, and metaphor Christian thought. That is, the central movements in Christian thought are substitutions—of spirit for letter, the atoning (substitutive) sacrifice of Jesus for man, the substitution of Jesus for the Oral Rabbinic Torah, etc. From a Rabbinic point of view, these substitutions are *distorted* displacements, an aspect of which is converting metonymy to metaphor, a displacement pretending to substitution wherein the metonymical distances between man and God, desire and fulfillment, subject and predicate are collapsed. These displacements rewrite and undo origins, usurp the Father, and break the metonymic chain.

The important point for the heretic hermeneutic, though, is that Rabbinic thought contains within itself the seeds, the potential for this heresy, through its own displacements. In fact, the very Rabbinic displacement of nature for Scripture is itself a kind of destruction of the logic of biological succession and filiation. It is commonplace to observe, for example, the theme of brotherly usurpation in the Bible, the disruption of the line of natural inheritance in the stories of Jacob and Esau, Isaac and Ishmael, and so forth. The Rabbis as well, in their role as brother interlocutors with God, accomplish, as we have seen, subtle displacements that enable them to achieve new identities and powers of the Text. The Jewish Son will discourse with God, will attempt his displacements through interpretation; the Christian Son will become God, resolve the tensions through the abolition of the metonymic discourse.

Displacement, usurpation, substitution. The problems of generation, succession, the "chain of signifiers," the belatedness of consciousness, and the interpreter are all central issues here. In Harold Bloom's poetics of belatedness, he postulates an "anxiety of influence" in which poets engage in a dialectical-historical struggle, each making room for himself by manipulating the tradition he has inherited, claiming to revise and purify it from error while in fact covertly overthrowing his predecessors. Through this revisionary interpretation, the poet can then see himself as his own father, redeemed from error, thus making the father a prefiguration of the son. Bloom's description of the poetic process helps illustrate aspects of the heretic Jewish hermeneutic I have been trying to articulate. This heretic hermeneutic is a complex of identification and displacement. Whether Moses comes to be read as a prefiguration of Jesus, as in Christian interpretation, or as a prefiguration of the heroic secular prophecy of Freud, in Freud's own identification with that Jewish leader, the dynamic is the same. The dead father in *Totem and Taboo* or *Moses and Monotheism* or *The Interpretation of Dreams*, or the Name-of-the-Father in Lacan, or all the Fathers of philosophy in Derrida—all enforce upon the living sons a certain contract of guilt and a desire to be free of that burden. And freedom comes in a characteristically Rabbinic mode—through interpretation, but here purposeful misinterpretation, or "misprision," to use Bloom's word. Under the guise of a new interpretation, the metonymic chain of succession is breached.

We might say that pious Rabbinic interpretation means the succession of links on the chain of metonymic signification, a chain where signified and signifier do not merge; one approaches the other only by interpretive approximation, and identity is constituted by positioning along this chain within the play of its polysemy. Heretic Rabbinic interpretation, as in Christianity, for example, is interpretation as breakage of the successive links, and replacement via metaphoric substitution, a union of signifier and signified, same and other, desire and fulfillment, the word become flesh.

On the other hand, the Rabbinic brother-interlocuter's displacements are also a dialectical identification with the Father. In Derrida's case, the heretic hermeneutic, while it is a displaced Rabbinism, is also at the same time a return to an identification with a kind of Jewish paternity that for so long had been opposed to and was repressed (and displaced) by the Greco-Christian tradition . . . and in Derrida's passionate defense of *Écriture*, we can begin to see it.

THE WANDERING OUTCAST OF LINGUISTICS

Barbara Johnson poses an important question in her perceptive critique of Derrida. To be honest, she says, Derrida's infinite play of signifiers would require him to play beyond the seme of Writing. Isn't Derrida, in effect, transforming "writing" into "the written," she asks?[25] Derrida's choice of Writing to oppose to Western logocentrism is, I am contending, a re-emergence of Rabbinic hermeneutics in a displaced way. Derrida will undo Greco-Christian theology and move us back from ontology to grammatology, from Being to Text, from Logos to *Écriture*—Scripture.

He will defend the Writing that has been viciously condemned and despised, and the letter that has been castigated as a carrier of death. We have discussed this stigma in Christian polemics against the Jews, but Derrida traces it from Greek philosophy to Rousseau, Saussure, Husserl, and even Lévi-Strauss. The fallen, literal dead writing is contrasted to a natural, living, venerated metaphoric writing—for example, the "voice of conscience" as divine law, or the writing of the heart, etc. Such a natural writing, however, is not grammatological but pneumatological.[26] The contrast of the writing of the soul/body, interior/exterior, consciousness/passions, and so forth dictates that one must return to the "voice of Nature" that merges with the divine inscription.

Derrida shows the persistence of this "onto-theology" even in secular thinkers. Writing is seen not merely as exterior, but as a threatening exterior from which spoken language must be protected, a corrupt menace that can erupt and disrupt the self-enclosed interiority of the soul (and Holy Family, as we have seen). In his discussion of "Linguistics and Grammatology" in *Of Grammatology*, Derrida maintains that Saussure sees writing as "perversion and debauchery, a dress of corruption and disguise, a festival mask that must be exorcised, that is to say warded off, by the good word," even as "original sin." Derrida is a vigorous polemicist who is adept at contorting the arguments of others to fit his own needs, but his particular use of passages and adjectives to characterize negative attitudes towards writing is somewhat curious: "the perverse cult of the letter-image," "the sin of idolatry," "perversion that engenders monsters," "deviation from nature," "principle of death," "deformation, sacrilege, crime," "the wandering outcast of linguistics," "expatriated, condemned to wandering and blindness, to mourning," "expelled other." The descriptions are overtly theological, and if logos comes out on the side of the "historical violence of a speech dreaming its full self-presence, living itself as its own resumption. . . . auto-production of a speech declared alive . . . a logos that believes itself to be its own father, being lifted above written discourse,"[27] it is obviously the Christian logos, the son dreaming himself to be his own father, born into the flesh and elevated above all texts and written discourses—and that exiled, wandering, mourning, condemned outcast accused of unredeemed original sin is the Jew, the carrier of the letter, the cultist of Writing.

It is odd that Derrida does not mention this most obvious point, especially since he is so much at pains to uncover the theological assumptions of this privilege of the logos. In essays written earlier than or contemporaneously with *Of Grammatology* and collected in *Writing and Difference*, Derrida quite clearly makes the connection between the Jewish/Christian polemic and the history of philosophy. We have already mentioned the essay on Edmond Jabès. Derrida discusses this question also in a long, admiring piece, "Violence and Metaphysics," on the French-Jewish philosopher Emmanuel Levinas, who so influenced him, and in "Ellipsis," which is the final essay of *Writing and Difference* and was specifically written for the book. In that final piece, Derrida argues for a "negative atheology," a writing that transcends the closure of "the book" for the openness of "the text." He signs the essay "Reb Derissa." Thereby his signature becomes the last words of the entire book. The change from Derrida to Derissa, from "-ridda" to "-rissa" in French could be one of those plays with words and on his name of which Derrida is so fond. If we take *risée* in French to mean "laughing or laughable," Derrida might be trying to have the last laugh on us as "Reb," or Rabbi, Derrida in an elliptical Rabbinic-commentary text-play.

But before we let him have the last laugh, which so many contemporary critics seem all too willing to do, let us take him a little more at his word. Derrida, of course, would object, since his very theory of free-play, part of his inheritance from Nietzsche, demands that we do not take him at his word, for there is no "proper," single, absolutely "true" meaning, and thereby Derrida can evade any theoretical attack and again assert his mastery over all critics. Derrida's deconstructive readings usually proceed by showing how the writer under consideration commits the very errors

of which he accuses others. Derrida thereby deconstructs, or undoes, the
writer's own theoretical foundations. In Derrida, too, I maintain, there is
an unacknowledged displaced theology, a displaced Rabbinism used both
to undo Greco-Christian culture, and to identify negatively and dialectically with a displaced and repressed Jewish hermeneutic.

So let us take Reb Derissa, the laughing rabbi, more in the spirit of
Midrash than Nietzsche, viewing his play as "serious play," as a commentary that is an extension of the text. As Gayatri Spivak writes, one of the
distinguishing characteristics of Derrida's method, which he inherits from
Freud, is an interpretive method that pays attention to the minute details
of a text, to syntax, the shapes of words—the dream's treatment of words
as things.[28] This is in effect a species of *Midrashic* play which makes Derrida's deconstructions so different from Heidegger's or Nietzsche's "de-
structions." Geoffrey Hartman writes:

> Derrida lets language be, not by nonchalance but by giving it its "to be," as
> he deconstructs a text or moves within, rather than simply against, equivo-
> cation and the multiple register of words. . . . Let no one mistake this
> nonbook: *Glas* is of the House of Galilee.
>
> Who else but Reb Derissa could go from the dissemination/castration or
> flower/sword theme to the Wartburg dictionary [in *Glas*] (59 ff.), and by an
> error . . . as bewildering as any semanticist has traced, show how sword
> and lily lie together *(lis/lit)* in 'glaïeuil' (gladiolus: *Schwertlilie*) with its
> *Blütenstaub* of phonic or dialectic resonances: *glageuil* ("klage," "deuil"?);
> *glaudius, claudio, gaudio* ("joy"?); *glaviol* ("viol"?); *glaive, glai, englasi*, ("ter-
> rify," "freeze," "glaze"?); *glai, glace, glisser?*[29]

ATHENS AND JERUSALEM

In a more serious moment, laughing Reb Derissa asks at the end of his
essay on Levinas:

> Are we Jews? Are we Greeks? We live in the difference between the Jew
> and the Greek, which is perhaps the unity of what is called history. We live
> in and of difference, that is, in *hypocrisy*. . . .
>
> Are we Greeks? Are we Jews? But who, we? Are we (not a chronolog-
> ical, but a pre-logical question) *first* Jews or *first* Greeks? And does the
> strange dialogue between the Jew and the Greek, peace itself, have the form
> of the absolute, speculative logic of Hegel, the living logic which *reconciles*
> formal tautology and empirical heterology. . . ? Or, on the contrary, does
> this peace have the form of infinite separation and of the unthinkable,
> unsayable transcendence of the other? To what horizon of peace does the
> language which asks this question belong? From whence does it draw the
> energy of its question? Can it account for the historical *coupling* of Judaism
> and Hellenism? And what is the legitimacy, what is the meaning of the

copula in this proposition from perhaps the most Hegelian of modern novelists: "Jewgreek is greekjew. Extremes meet"?[30]

The history of philosophy is ultimately an argument between Jews and Greeks. Levinas, in his many influential philosophical works, has taken up the side of the Jews. The influence of Levinas on Derrida is not often mentioned. Derrida openly acknowledges it in his essay on Levinas, which begins with Matthew Arnold's famous quotation from *Culture and Anarchy*, about Hebraism and Hellenism being the two major forces of culture. Philosophy, Derrida points out, is Greek in the most ethnocentric sense, and even Husserl and Heidegger, who seek to subordinate and transgress this metaphysical tradition, are not free from its Greek elements. "Here," writes Derrida, "the thought of Levinas can make us tremble. At the heart of the desert, in the growing wasteland, this thought which fundamentally seeks to be a thought of Being and phenomenality makes us dream of an inconceivable process of dismantling and dispossession." In Levinas, Derrida finds an attempt to "dislocate the Greek logos," and thus to dislocate our identity and the principle of identity in general, which is a summons to depart from Greece, to liberate thought from the "oppression" of the Same and the One, an "ontological or transcendental oppression," which Derrida claims is "the origin or alibi of all oppression in the world.[31]

With Levinas we have a "parricide of the Greek father Parmenides," as Derrida puts it, the theme of murder of the primal father and usurpation of the origins again resurfacing here. Derridean deconstructionism, too, will murder the father-founders of philosophy and disseminate a new Writing, which, in the wake of the overthrow of the Same and the One, celebrates pluralism, otherness, distance, and difference. Parmenides' disregard of the Other is "totalitarian" and tautologous, according to Levinas and Derrida, and the rebellion against the Greeks is a species of liberation.

In place of presence, or *ousia*, Derrida will put his Writing, which is "more 'metaphysical' than speech." The writer can absent himself better, address himself more effectively to the other, can better defer, delay, multiply signs, and renounce the immediacy of violence. And Levinas teaches Derrida that Hebraism in its connection with the letter has much to show us. Derrida quotes three passages from Levinas:

> "To admit the action of literature on men—this is perhaps the ultimate wisdom of the West, in which the people of the Book will be recognized."

> "The spirit is free in the letter, and subjugated in the root."

> "To love the Torah more than God [is] protection against the madness of a direct contact with the Sacred."[32]

The latter quotation Derrida takes from Levinas's *Difficile Liberté: Essai sur le Judaism*, which is worth examining somewhat more closely to trace the Judaic sources of some of Derrida's ideas. "To love the Torah more than God" is the title of the essay, a title which has its source in the famous statement found in Midrash and Talmud that the rabbis place in the mouth of God: "So should it be that you would forsake me, but would keep my Torah" (Yer. *Hagigah* 1:7, Lam. *Ber. Rab.*, intro. ch. 2). The statement is striking and eminently Rabbinic—the Torah, the Law, Scripture, God says, are even more important than He. We might say that Derrida and the Jewish heretic hermeneutic do precisely that: forsake God but perpetuate a Torah, Scripture, or Law in their own displaced and ambivalent way. Derrida, above all, keeps faith with the poor exiled scapegoat, "Writing."

Levinas, it should be noted, writes a specifically post-Holocaust French-Jewish philosophy. He is himself a survivor of the camps, and the experience of the absence of God is the main concern of his essay on the Torah and God. The absent God of the Holocaust, the God who obscures His face, paradoxically becomes for Levinas the condition of Jewish belief. The loss of a consolatory childish heaven, the moment when God withdraws from the world, is the moment that calls for what Levinas describes as an "adult" faith, where the adult can triumph only in his own conscience and suffering, a suffering which is no "mystic expiation of the sins of the world" but an ordeal of an adult, responsible man, "a suffering of the just for a justice without triumph, [which] is lived as Judaism." The relation of man and God in Jerusalem is

> not a sentimental communion in the love of an incarnate God, but a relation between spirits, through the intermediary of a teaching, the Torah. It is precisely a discourse, not embodied in God, that assures us of a living God among us. . . . [To love the Torah more than God is] protection against the madness of direct contact with the Sacred without the mediation of reason.
> . . . The spiritual does not present itself as a tenable substance but, rather, through its absence; God is made real, not through incarnation, but, rather, through the Law.[33]

Judaism is then defined as this trust in an absent God.

In his commentary on this essay, Richard Sugarman makes the important point that in Levinas absent justice does not mean that jusitce is nonexistent. Sugarman goes on to say:

> This decisive metaphysical distinction between the phenomenon of absence and that of non-existence, so long obscured in the history of philosophy, is central to Levinas' analysis and needs to be made more explicit. Absence is

not that which is merely somewhere else, convertible into presence by a change of position, perspective, or interpretation. That which is absent is not necessarily an entity in another place, hidden from view, or unintelligible. Rather, the phenomenon of absence positively informs our understanding of everyday events with considerable concrete significance.[34]

This point is crucial to understanding the schism between Jews and Greeks, and between Derrida and the history of philosophy: *absence does not equal nonexistence.* Absence, otherness, the "trace," all of Derrida's prime terms, are part of a vocabulary that seeks to evade the trap of Being or Nonbeing of Greek philosophy. Derrida's reality is not Being, but Absence; not the One, but the Other; not Unity, but plurality, dissemination, writing, and difference. Reb Derissa, too, claims that "the spiritual does not present itself as a tenable substance but, rather, through its absence," and he is the new High Priest of the religion of Absence.

Derrida agrees with Levinas that hearing is higher than seeing, for seeing the face is presence, *ousia.* And it is in discourse with God, the absent Other, that Levinas founds his "Rabbinic" metaphysics. Derrida writes in his commentary:

> Via the passageway of his resemblance, man's speech can be lifted up toward God, an almost unheard of *analogy* which is the very moment of Levinas's discourse on discourse. Analogy as dialogue with God: "Discourse is discourse with God. . . . Metaphysics is the essence of this language with God." Discourse with God, and not in God as *participation.* Discourse with God, and not discourse on God and his attributes as *theology.* And the dissymmetry of my relation to the other, this "curvature of inter-subjective space signifies the divine intention of all truth." It "is, perhaps, the very presence of God." Presence as separation, presence-absence—again the break with Parmenides, Spinoza and Hegel, which only "the idea of creation *ex nihilo*" can consummate. Presence as separation, presence-absence as resemblance, but a resemblance which is not the "ontological mark" of the worker imprinted on his product, or on "beings created in his image and resemblance" (Malebranche); a resemblance which can be understood neither in terms of communion or knowledge, nor in terms of participation and incarnation. A resemblance which is neither a sign nor an effect of God . . . "the trace of God."[35]

Derrida's analysis of Levinas is also an apt summary of the trends of Rabbinic thought that we have discussed: a hermeneutic developed independently of ontology; a text whose writing is precisely this presence-as-separation, and which is structured not around a hierarchical great chain of being, but which conceives of metaphysics as a discourse with God, an endless dialogue and disputation, interpretation and re-interpretation.

Christianity, on the other hand, is bound to a Greek ontology, in which beings are related to Being via participation, not discourse; incarnation, not interpretation; and in which absence is intolerable. In the reconciliation of the logos with itself, what is other must be returned to the same. If Judaism is the experience of the infinitely other, it is precisely this irruption of the totally other that threatens the Greek logos—and the Christian Holy Family as well. Like Writing, the Jew is historically the castigated other, intruder, threat, scapegoat, exile, idolater, and it is surely no accident that those who take up arms against the paternal logocentrism of Western thought should be Jews. Though secularized, though having forsaken the Jewish God, they "keep the Scripture."

But Derrida would go further than Levinas. He will invert and displace the Rabbinic hermeneutic. What if the world is not the effect of the "trace of God," but the reverse—if "God is the effect of the trace"? In *Of Grammatology*, Derrida acknowledges his debt to Levinas for the relation of the concept of the trace to the critique of ontology.[36] In replacing ontology and semiology with grammatology, Derrida claims that "difference" precedes all similarity; and the trace, which is pure difference, does not "exist" as a presence outside of plenitude, but is the condition of that plenitude and anterior to all signs: *"The trace is in fact the absolute origin of sense in general. Which amounts to saying once again that there is no absolute origin of sense in general. The trace is the differance* which opens appearance and signification" (p. 65). The language of Derrida here is, to say the least, a kind of mystification, an attempt to transcend a system of thought he discredits by using his own tools, in a species of *via negativa*. The "origin" is this "trace," a presence-absence that "carries in itself the problems of the letter and the spirit, of body and soul, and of all the problems whose primary affinity I have recalled" (p. 71). The entire history of metaphysics, Derrida claims, with all its dualisms and monisms, has striven to reduce the trace and subordinate it to the full presence of the logos, and thereby humble writing. This onto-theology determined the meaning of being as presence, parousia, life without difference, from Plato to infinitist metaphysics:

> We must not therefore speak of a "theological prejudice," functioning sporadically when it is a question of the plenitude of the logos; the logos as the sublimation of the trace is *theological.* (p. 71)

Harold Bloom, always one to spot a writer's revisionary blindness to his precursors, catches Derrida here. Bloom asserts that "the trace," "Writing," "difference" indeed all appear in the Jewish mystical tradition—in Kabbalistic interpretation:

> Derrida says that "all Occidental methods of analysis, explication, reading
> or interpretation" were produced "without ever posing the radical question
> of writing," but this is not true of Kabbalah. . . . Kabbalah too thinks in
> ways not permitted by Western metaphysics, since its God is at once *Ein-*
> *Sof* and *ayin*, total presence and total absence, and all its interiors contain
> exteriors, while all of its effects determine its causes.[37]

Moreover, Bloom perceives that despite Derrida's posture as archdecon-
structionist and playful nihilist, he is in fact performing Rabbinic revisions, Jewish corrections of Western philosophy:

> Though he nowhere says so, it may be that Derrida is substituting *davhar*
> for logos, thus correcting Plato by a Hebraic equating of the writing-act
> and mark-of-articulation with the word itself. Much of Derrida is in the
> spirit of the great Kabbalistic interpreters of Torah, interpreters who create
> baroque mythologies out of those elements in Scripture that appear least
> homogeneous in the sacred text.[38]

REB DERISSA, KABBALIST

Bloom prefers to see Derrida not as "Rabbinic" but as Kabbalistic, and
would make a major distinction between Kabbalah and normative orthodox Judaism. For Bloom, Kabbalah is a strong misreading of the orthodox canon, a model for poetry and criticism because it forcefully
manipulates, opens, misreads, revises the tradition in accordance with its
own catastrophic vision. He interprets Kabbalah as the Jewish response to
exile in the extreme—a passionate opening of the Sacred Text to the
sorrows of time and history. Bloom is actually reading/misreading Gershom Scholem's masterful studies of Kabbalah,[39] for Scholem has argued
that Kabbalah is the extreme extension of the Rabbinic freedom with the
text that we have discussed previously—the attempt to include within
revelation all the later tradition and commentary that would be offered to
explain its meaning.

As we have noted, carried to the extreme, or under the pressure of
historical or other catastrophe, such free revisionist reading can become
antithetical. A disastrous example among Jewry was the false and heretical
messianic movement of Shabbatai Zevi in the seventeenth century.
Scholem, however, presents Kabbalah and its varying distorted mani-
festations—which had long been held in disrepute by the nineteenth century German rationalist scholars—as the very vivifying heart of Judaism.
And Scholem and Kabbalah can help us to understand Derrida's specific
heretic hermeneutic.

As David Biale shows in his study of Gershom Scholem, Kabbalah,

Gnosticism, and Sabbatianism became "a powerful weapon for Scholem in shattering dogmatic definitions of Judaism by showing how censored 'heresies' in Jewish history were just as legitimate as the normative tradition. The argument that the Sabbatian messianic heretics were part of Jewish history became the cornerstone of his counter-history."[40] Scholem's critics, however, saw in this revision of Jewish history a manifestation of Scholem's own antinomianism, an attempt to disguise his own subversion by re-interpreting the tradition itself as subversive, and a project to make his own secular interpretation of Judaism part of the normative Jewish tradition.

If we disregard the judgmental aspect of these critics, and take their statements as descriptive rather than prescriptive, I think they are extremely suggestive. In interpreting "heretical" as truly a "normative" part of Jewish history, Scholem's "antinomianism" is here a *return to* tradition. And we can say that the "heretical" impulses of Freud, Derrida, and Bloom are likewise a dialectical return to Jewish tradition. Scholem's heretic hermeneutic results not in a break from tradition, but in a vision of heresy as deeply traditional; and he demonstrates this through a special kind of interpretive commentary on Scripture. What could be more traditional? The paradox is the return to tradition by way of heresy. And this movement is precisely what Scholem locates as the center of Kabbalah, which created, in his words, an "Orthodox Gnosticism"—an oxymoronic term that well expresses this paradox.

Scholem's stance towards his precursors, Biale shows, is modelled after what Scholem conceived to be the stance of Kabbalah toward the Scriptural tradition of "normative" Judaism: Kabbalah itself was a "counter-history" within Judaism which appropriated and transformed it by integrating mythic, Gnostic, and irrational mystical concepts into the heart of monotheism, and which was thereby able to revitalize it.

Scholem as counter-historian of Kabbalah through his secular historical method seeks also to place himself within Jewish tradition and thereby transform it by revealing the heresy at the heart of this tradition—thus developing a heretic hermeneutic that claims to revitalize even as it abrogates tradition. Scholem investigates what had been consigned to the "cellar" of Jewish history—the subterranean, suppressed, subversive, esoteric tradition that had run counter to the official version of Judaism created by the historians and the rabbis. But in this cellar one finds the secret of Jewish survival. The parallels to Freud, Derrida, and Bloom are obvious: they also play the role of outsiders to their respective "traditions"—scientific, philosophic, literary. They, too, look to the "cellars" of self, existence, poetry, seeking there suppressed hidden movements which are "heresy" to the "normative" views of self, of Being, of literature. They, too, find irrational, demonic, mythic secrets in the heart of reason, science, humanist tradition. In uncovering these secret "anarchic" forces, they subvert what had been the normative "orthodox" view and make the outside inside. "Heresy" becomes "tradition," and they attain a priority and authority over those traditions to which they were heirs. Derrida undoes the very origins of Western philosophy; Freud uncovers the heretical desires at the origins of the self; and Bloom destroys the gentilized tradition of Eliot *et al.* by uncovering the fierce Oedipal warfare at the heart of poetry. All of these acts are seizures of the original Texts of tradition, inversions of them through feats of interpretation, and at the same time affirmations of fidelity to the "secret" tradition, which then becomes the "real" tradition; and these masterly interpreters become the Moses figures who bring the revelation to the people from the flaming mountain. These interpretive mediations—however secular—become the only "revelation" possible.

MYSTICAL TRACES

In this shadowy area, interpretation "crosses the line," revisionism becomes heresy, and heresy tradition. The heretic hermeneutic in effect continues, even as it attempts to overcome tradition. The point at which revisionism becomes heresy, though, is subtle. Scholem finds the key in the Kabbalists' intense concentration on the nature of the divine language in their attempts to penetrate the inner meaning of the Divine Text. Bloom in fact reads Kabbalah precisely as a theory of rhetoric and perceptively marks the affinity between the Kabbalistic view of language and the Derridean:

> Language, in relation to poetry, can be conceived in two valid ways, as I have learned, slowly and reluctantly. Either one can believe in a magical theory of all language, as the Kabbalists, many poets, and Walter Benjamin did, or else one must yield to a thoroughgoing linguistic nihilism, which in its most refined form is the new mode now called Deconstruction. But these two ways turn into one another at their outward limits. . . . Is there a difference between an *absolute* randomness of language and the Kabbalistic magical absolute, in which language is totally over-determined?[41]

Indeed, Derrida's notion of the trace sounds strangely similar to Scholem's description of the Kabbalistic mystical "Name of God." Relevant in Scholem's analysis is the Kabbalistic concept that the divine language of the Written Torah was itself *already mediated*—and that the essence of the divine language was the mystical "Name of God" which was encoded in the text. Thus the actual text of Scripture is thought to be composed of the various combinations and permutations of this mystical Name.[42]

Scholem understands this Name of God to be somehow equivalent to His essence, and as an emanation and creative power transcending any human language, grammar, or understanding. Itself "above or beyond" meaning—"meaningless" as Scholem puts it—this mystical Name is nevertheless the inexhaustible source of all meaning, and thus opens out into infinite interpretation: "This absolute word is originally communicated in its limitless fullness, but—and this is the key point—this communication is incomprehensible!"[43] It becomes comprehensible only as it is mediated through the interpretations of tradition.

Scholem makes a radical distinction between this mystical "meaningless" word and the words of Scripture and tradition which interpret it; for who can know the meaning of the ultimately "meaningless" word? The radical consequence which, according to Scholem, the Kabbalists veil, but which he claims to reveal, is finally that there is "no such thing as Written Torah in the sense of an immediate revelation of the divine words." In the following passage one could substitute "text" for "Written Torah," and "interpretation" for "Oral Torah," and it would sound quite Derridean:

> The Written Torah is itself mediated; there is thus only and already interpretation, only Oral Torah. This Oral Torah, however, still retains the character of the absolute, and bears the process of infinite interpretation. As opposed to the idea of revelation as a specific communication, revelation which has yet no specific meaning, is that in the word which gives an infinite wealth of meaning. Itself without meaning, it is the very essence of interpretability. For mystical theology, this is a decisive criterion of revelation.[44]

And so it seems the decisive criterion for Derrida as well—an "original" meaningless trace which is nevertheless the source of an infinite interpretive play.

Another passage, in the *Zohar*, one of the central Kabbalistic works describing the process of creation, contains a description of a mystical phenomenon strikingly similar to Derrida's trace:

> When the most Mysterious wished to reveal Himself, He first produced a single point which was transmuted into a thought, and in this He executed innumerable designs, and engraved innumerable engravings. He further graved within the sacred and mystic lamp a mystic and most holy design, which was a wondrous edifice issuing from the midst of thought. This is called MI [Who?], and was the beginning of the edifice, existent and non-existent, deep-buried, unknowable by name. (*Zohar* 1b)

But is not the reinstitution of the trace then also theological? one may ask Derrida. Is not the "trace" dialectically related somehow to all it negates? Well aware of these objections, Derrida tries to warn against any "return to the Book" in his discussion of Jabès in the essay "Ellipsis," at the end of *Writing and Difference*. The Book must finally be closed and the Text opened, for "if closure is not end, we protest or deconstruct in vain," says Derrida, echoing Paul's "If Jesus is not risen, our faith is in vain." The echo of Paul is most revealing; it is a variation on Jewish heretic hermeneutic. This time, however, it is not something so crass as an incarnate god that will supplant Scripture, but something more subtle and far more Jewish—a Text, a "writing beyond Book." This writing "feigns," by repeating the book, inclusion in the book, but does not let itself be enveloped within the volume. It is "the writing of the origin, the writing that retraces the origin, tracking down signs of its disappearance, the lost writing of the origin."[45] What takes the place of the origin is not absence but a trace, that is, origin by means of which nothing has begun. And thus another slaying of Moses, dispossession of the father and re-appropriation of the father and origin, a new Writing. But this time one that seeks to secure itself from all future displacement by concocting a mystifying non-terminology that at every instant eludes definition.

PLAYING IN EXILE

Derrida's specific form of Jewish heresy is not metonymy become metaphor but metonymy run amok, metonymy declaring itself to be independent of all foundations and yet claiming to be the origin and law of everything. *Différance* means in Derrida to differ and to defer, or postpone. Whereas Levinas claims that at some point there must be an unveiling, a redemption however long postponed, Derrida chooses to stay in Exile, to infinitely defer and differ—to play. Derrida will play in the interval between Book and Book, play with the "center," and de-center origin through writing. He will define the center as only a hole; writing the hole, "we plunge into the horizontality of a pure surface, which itself represents itself from detour to detour." Derrida will have the last laugh on all pretenders to his new throne of Writing. No one can slay this new Moses. At any point, he can take another detour, or choose to differ. Derrida's feigning the book is also a feigning of philosophy, feigning seriousness, feigning play.

Derrida is the prodigal son, but unrepentant, enjoying his escapade. Geoffrey Hartman notes this theme in Derrida's theory of "dissemination," the play with "that which does not return to the father.". . . . It is a word cast on the waters, a prodigal without hope of return. The "imitation of nature" now takes nature literally and substitutes the image of a creative self-scattering for the "collected" imitation of a divine pattern: the "legein" of the logos.[46]

This logos cannot return to the Father, and no text can return to its author. Thus Derrida, too, is free from any sins that his own prodigal texts might commit. He is secure from all attack and reproach. This new setting aside of the father frees him from the original sin to which Freud was so grimly tied in his own theory of parricide. Derrida will joyfully carry out his "cruci-fiction of the Word," as Hartman puts it, with no remorse. He will remain in Exile, and exile the logos wih him. He would not come home and he would not welcome the Messiah.

If psychoanalysis might be seen as one attempt to cure the neurosis of the Jew-in-Exile, deconstructionism could be thought of as another. Kafka made the following perceptive comment about the exilic character of psychoanalysis:

> It is no pleasure to busy oneself with psychoanalysis, and I keep as far away from it as possible, but it has at least as much reality as this generation. The Jews have always produced their joys and sorrows at almost the same time as the Rashi commentary relating to them, and here again they have done so.[47]

Marthe Robert comments on this remark that Kafka finds the meaning of Freud in the context of the present-day joys and sorrows of Jewish life, which is unique in that "it is obliged since time immemorial to provide its own commentary, it has always been a written life, produced not before but almost at the same time as the writings that explain it."[48]

Let the commentary, then, says Derrida, the Writing developed in the endless delay of Exile, be all, and be playful. Let Exile subvert Being and Logos entirely. "Encounter is separation," Derrida writes, echoing Levinas, a proposition that "breaks the unity of Being. . . . by welcoming the other and difference into the source of meaning."[49] This

> original exile from the kingdom of Being, signifies exile as the conceptualization of Being, and signifies that Being never is, never shows *itself*, is never *present*, is never *now*, outside difference. . . . Whether he is Being or the master of beings, God himself is, and appears as what he is, within difference, that is to say, as difference and within dissimulation. [p. 74]

This difference within God is what Derrida refers to in the essay on Jabès as the Broken Tablets of the Law, which allow for poetic autonomy and the freedom of man's speech:

> Between the fragments of the broken Tables the poem grows and the right
> to speech takes root. Once more begins the adventure of text as weed, as
> outlaw far from *"the fatherland of the Jews,"* which is a *"sacred text surrounded
> by commentaries."* [p. 67]

Both the poet and the Jew must write and must comment, because both
poetry and commentary are forms of exiled speech, but the poet need not
be faithful nor bound to any original text. For Jabès, the Law, after the
breaking of the Tablets, becomes a question, and to interrogate becomes a
duty. The Broken Tablets represent a negativity and difference in God, a
dissimulation and hiding of His Face, which Derrida claims is the origin
of our freedom—and our Writing. Derrida is here attracted to the same
theme as found in Levinas—the Judaic sense of the absence of God as His
presence, and God's duplicity, his hiding and obliqueness. (And he recognizes that the themes of "the question within God . . . [of] Negativity in
God, exile as writing, the life of the letter are all already in the Kabbalah"
[p. 74].) Kabbalah speaks, for example, of God's self-contraction, or *tzimt-
zum*, as the originating movement of creation.) God's detours are borrowed by man; this infinite detour, or "Way of God," is "preceded by no
truth, and thus lacking the prescription of truth's rigor, is the way through
the Desert. Writing is the moment of the desert as the moment of Separation" (p. 68).

Here Derrida, too, takes up residence and pitches his tent. Between the
fragments of the Tablets he has destroyed, his texts grow up like "weeds,"
"outlaws," imitating the ruses and absences and infinite detours of a hiding God. Derrida wants to flourish in the absent spaces *between* the Tablets, *between* Jew and Greek, Rabbi and Poet. Absence, Derrida claims, is
the "letter's ether and respiration"; signification arises through absence,
rupture, fragmentation, the discontinuity of the letter:

> The caesura makes meaning emerge. It does not do so alone, of course; but
> without interruption—between letters, words, sentences, books—no
> signification could be awakened. *Assuming* that Nature refuses the *leap*, one
> can understand why Scripture will never be Nature. It proceeds by leaps
> alone. Which makes it perilous. Death strolls between the letters. To write,
> what is called writing, assumes an access to the mind through having the
> courage to lose one's life, to die away from nature. (p. 71)

For Derrida, absence is the ground and the *content* of the letter. This
wandering life of the letter expresses itself above all in metaphor,
metaphor as the origin of language, beyond Being and Nothing:
"Metaphor, or the animality of the letter, is the primary and infinite
equivocality of the signifier as Life. The *psychic* subversion of inert literal-

ity, that is to say, of nature, or of speech returned to nature" (p. 73). This is a mode of metaphor which does not return to its "natural" or "proper" univocal meaning, but subverts nature. Such a metaphorical letter or word could never become flesh incarnate, and will not "return speech to nature," as does the Christian logos. Derrida will yet wander among the infinite play of letters of Scripture, between the lines. He is more at home in Scripture than Nature, as are the Rabbis.

But Derrida, as we have seen, cannot reconcile himself to remaining somehow within the Book; he would pose a final question to Jabès's *Book of Questions*, a question that would finally free Derrida from the Epoch of the Book and make it no longer the model of meaning. Instead, he will ask if the meaning of Being is not a "radical illegibility" (p. 77). This radical illegibility of an era other than the Book is not, Derrida claims, "irrational," or something that is defined in relation to logic and the Book. It is, rather, prior, an "original illegibility" that is the very possibility of the Book. "The Being that is announced within the illegible is beyond these categories, beyond, as it writes itself, its own name" (p. 77). (Again this sounds strikingly like the Kabbalistic unintelligible mystical Divine Name or "point" preceding creation.)

Again the passion to displace origins, the Jewish heretic hermeneutic, surfaces in Derrida, who places a radical unintelligibility at the origin of his thought, a radical illegibility that constitutes, in fact, his own work. He, too, tries to write beyond the finitude of his own "proper name" with his fantasy of a disseminated name, which he places at the origin. Like Freud, he has his own species of family romance within the family of philosophy, where he can subvert his Judaeo-Greek origins and evade any attempt to catch him by taking up residence in the absent spaces between his lines. Derrida will always choose to differ, but he leaves no empty spaces for any others who would differ with him.

He has both foreclosed the history of philosophy and even appropriated the blanks, the nonmeanings, even the space of illegibility. Even here, however, he will dissimulate in order to disseminate himself. The ultimate questions he asks of the Book must, he says, sleep. "Writing would die of the pure vigilance of the question, as it would of the simple erasure of the question. Is not to write, once more, to confuse ontology and grammatology?" (p. 78). Derrida has too much to lose if Writing were to die. He must defer his own questions, or else his radical attack on origins and the Book would lead to his own dissolution and silence. Derrida must somehow perpetuate the Law, at least be the father of Writing lest the parricide become an inadvertent suicide.

The Jewish prodigal sons cannot completely forgo Scripture. To attack the European psyche and the Holy Logos, to attempt to overturn Western

man from within and without is an act of revenge by the exiles—and yet
again a defense of the Jewish father. They will try, nevertheless, to recap-
ture the "purloined letter," to redeem Scripture from the abuses it has
suffered at the hands of Greeks and Christians. And they will accomplish
their victory, above all, through acts of interpretation.

There is for them, however, no end to exile. There is no fulfillment of
signs, but rather a raising of the Jewish condition of exile into a paradigm
of existence: to be is to be in exile; to make texts is to already interpret;
absence is presence. Or, perhaps, in reaction to a Scripture that endlessly
promises but never fulfills, they will make of their exile an antithetical
Promised Land, a "Criticism in the Wilderness" to use the title of one of
Geoffrey Hartman's recent books.

How are we to understand, finally, this complicated interweaving of
tradition, revision, and heresy? Only the language of parable, at last, can
capture it. Kafka, who so painfully shared these anxieties of influence and
secular-religious dilemmas, tells the story in a chilling parable:

> Leopards break into the temple and drink the sacrificial chalices dry; this
> occurs repeatedly, again and again: finally it can be reckoned on beforehand
> and becomes part of the ceremony.[50]

Kafka's parable seems to me uncanny in its description of Freud, Derrida, Bloom, and the Jewish heretic hermeneutic. The Jewish Temple in
fact was broken into several times and destroyed twice; its devotees were
sent into exile, displaced yet forever yearning to return. In the long in-
terim of exile, study and interpretation of the laws and rituals of the
Temple replace the literal sacrifices. In exile, a broken people try to heal
themselves through ever more complicated figuration, opening, troping of
their Sacred Text, trying somehow to make the facts of their historical
catastrophe agree with the exalted promises of their Sacred Book. And
this can be accomplished only through feats of subtle interpretive reversal:
somehow the leopards have entered the Temple and must be accom-
modated without being allowed to triumph. Excessive troping, transgressive interpretation, Kabbalistic inversion, and displacement all appear
under the guise of extension and application of the Sacred Book, part of its
unfolding interpretation. Bloom is so attracted to Kabbalah as a paradigm
for all revisionist thought and model for intrapoetic wars precisely for this
reason; he recognizes the excesses of Rabbinic interpretation as survival
strategies, "necessary misreading" not only against historical catastrophe,
but against an overwhelming, authoritative paternal Sacred Text that can
never be overcome.

The leopards, the rebellious, profaning, heretical forces have become part of the holy ritual itself—through interpretation. "Displacement" not only describes such extreme misreading, is not only a technique of interpretation; it is the only way to survive the endless displacements of Jewish history. Displacement is a necessary re-vision and re-creation of a Text which is the only anchor of a people displaced in space. Displacement, in other words, is both the *condition* and *answer* to exile. The leopards become part of the ritual. In the school of heretic hermeneutics, holy and profane intermingle; there is something sacred about writing, commentary, and Texts, yet these notions are displaced into the profane fields of literature, philosophy, psychoanalysis. At the same time, under the guise of revisionary interpretation, the Temple and Scripture are profaned. The lines become crossed: who knows now which is the holy, and which the profane—which the leopards and which the priests?

NOTES

1. Jacques Derrida, *Glas* (Paris: Galilée, 1974), pp. 2686–69b, cited and trans. in Gayatri Spivak, "*Glas*-Piece: A Compte rendu," *Diacritics* 7 (1977): 23. "*Derrière les rideaux*" has encoded within it the syllables of Derrida's name: derri-da. *Glas* is full of this sort of self-inscribing wordplay.

2. Spivak, "*Glas*-Piece," p. 23.

3. Jacques Derrida, "Edmond Jabès and the Question of the Book," in *Writing and Difference*, trans. Alan Bass (1967; Chicago: University of Chicago Press, 1978), pp. 64–67.

4. Ibid., p. 65.

5. Bloom focuses on this central question in all of his latest theoretical works including: *The Anxiety of Influence* (New York: Oxford University Press, 1973); *A Map of Misreading* (New York: Oxford University Press, 1975); *Kabbalah and Criticism* (New York: Seabury Press, 1975); *Poetry and Repression* (New Haven: Yale University Press, 1976).

6. Simon Rawidowicz, "On Interpretation," in *Studies in Jewish Thought*, ed. Nahum Glatzer (Philadelphia: Jewish Publication Society, 1974), pp. 45–80. See also "Israel's Two Beginnings: The First and the Second 'Houses'" in the same volume, pp. 81–209.

7. Geoffrey Hartman, *Criticism in the Wilderness: The Study of Literature Today* (New Haven: Yale University Press, 1980), p. 202.

8. Ibid., p. 206.

9. Rawidowicz, "On Interpretation," p. 52.

10. Jacques Derrida, *Of Grammatology*, trans. Gayatri Spivak (1967; Baltimore: Johns Hopkins University Press, 1976), p. 13.

11. Ibid., p. 14.

12. On this general topic, see the studies by Thorlieff Boman, *Hebrew Thought Compared with Greek* (Philadelphia: Westminster Press, 1954); Isaac Rabinowitz, "'Word' and Literature in Ancient Israel," *New Literary History* 4 (1972):119–30; and the existentialist philosophical analysis by Lev Shestov, *Athens and Jerusalem*,

trans. Bernard Martin (Athens: Ohio University Press, 1966). See also the excellent study by G. Douglas Atkins, "Dehellenizing Literary Criticism," *College English* 41 (1980): 769–79, which delineates the anti-Hellenic and pro-Hebraic tendencies of the "Yale School" of literary critics—Geoffrey Hartman, Jacques Derrida, Harold Bloom, Paul de Man, and J. Hillis Miller.

13. Hans Jonas, *Philosophical Essays: From Ancient Creed to Technological Man* (Englewood, N.J.: Prentice-Hall, 1973), p. 29.

14. On this topic see Louis Jacobs, *Studies in Talmudic Logic and Methodology* (London: Vallentine, Mitchell, 1961); Tzvetan Todorov, "On Linguistic Symbolism," *New Literary History* 6 (1974): 11–34; Jacques Derrida, "The Supplement of Copula: Philosophy *before* Linguistics," in *Textual Strategies: Perspectives in Post-Structuralist Criticism*, ed. Josué V. Harari (Ithaca: Cornell University Press, 1979), pp. 82–120; and Susan Handelman, "Greek Philosophy and the Overcoming of the Word," *Works and Days* 1 (1980): 45–69.

15. See Derrida's long essay on this subject, "White Mythology: Metaphor in the Text of Philosophy," *New Literary History* 6 (1974): 5–74.

16. Paul Ricoeur, *The Rule of Metaphor*, trans. Robert Czerny (Toronto: University of Toronto Press, 1977), pp. 20–22.

17. John Freccero, "The Fig Tree and the Laurel: Petrarch's Poetics," *Diacritics* 5 (1975): 37.

18. For an extended Derridean reading of Augustine, see Margaret Ferguson, "St. Augustine's Region of Unlikeness: The Crossing of Exile and Language," *Georgia Review* 29 (1975): 842–64.

19. Quoted in Robert L. Wilken, *Judaism and the Early Christian Mind* (New Haven: Yale University Press, 1971), p. 1.

20. Jean-François Lyotard, "Jewish Oedipus," *Genre* 10 (1977): 401–3.

21. Lyotard, p. 406. He also writes: "What's in Hamlet that's not in Oedipus? There is non-fulfillment. This can be seen as the psychic dimension of neurosis or the tragic dimension of thought. It has quite another dimension. Oedipus fulfills his fate of desire; the fate of Hamlet is the non-fulfillment of desire; this chiasmus is the one that exists between what is Greek and what is Jewish, between the tragic and the ethical. . . . In Hebraic ethics, representation is forbidden, the eye closes, the ear opens in order to hear the father's spoken word. The image figure is rejected because of its fulfillment of desire and delusion; its function of truth is denied" (pp. 400–402).

22. Derrida, *Glas* 99a, trans. Spivak in "*Glas*-Piece," p. 33.

23. Derrida, *Glas* 67a, in Spivak, "*Glas*-Piece," p. 33.

24. Roman Jakobson, "Two Aspects of Language: Metaphor and Metonymy," in *European Literary Theory and Practice*, ed. Vernon Gras (New York: Delta, 1973), p. 121.

25. Barbara Johnson, "The Frame of Reference: Poe, Lacan, Derrida," *Yale French Studies* 55–56 (1977): 483.

26. Derrida, *Of Grammatology*, p. 17.

27. Ibid., pp. 35, 38–39, 44.

28. Spivak, Introduction to *Of Grammatology*, p. xlv.

29. Geoffrey Hartman, "Monsieur Texte," in *Saving the Text: Literature/Derrida/Philosophy* (Baltimore: Johns Hopkins University Press, 1981), p. 19.

30. Jacques Derrida, "Violence and Metaphysics: An Essay on the Thought of Emmanuel Levinas," in *Writing and Difference*, p. 153.

31. Derrida, "Violence and Metaphysics," pp. 82, 83. See also Lev Shestov's

Athens and Jerusalem,n. 12 above, a long meditation on this subject. A Russian Jew who taught philosophy in Paris, Shestov is probably one of Derrida's unacknowledged precursors.

32. Derrida, "Violence and Metaphysics," pp. 89, 102.

33. Emmanuel Levinas, "To Love the Torah more than God," trans. and rpt. Helen A. Stephenson and Richard Sugarman, *Judaism* 28 (1979): 218, 219. "*Difficile liberté: essai sur le Judaism*," in *Presences du Judaism* (Paris: Albin Michel, 1963), pp. 218, 219.

34. Sugarman, "Torah," p. 221.

35. Derrida, "Violence and Metaphysics," p. 108.

36. Derrida, *Of Grammatology*, p. 70. Succeeding references to the *Grammatology* are cited parenthetically in the text.

37. Bloom, *Kabbalah and Criticism*, p. 53.

38. Bloom, *A Map of Misreading*, p. 43.

39. Among Scholem's major works are: *Major Trends in Jewish Mysticism* (1941; rpt. New York: Schocken, 1961); *On the Kabbalah and its Symbolism* (1960; rpt. New York: Schocken, 1969); *Kabbalah* (New York: New American Library, 1974); *The Messianic Idea in Judaism and Other Essays on Jewish Spirituality* (New York: Schocken, 1971); *Shabbatai Zevi: The Mystical Messiah, 1626–1676*, trans. R. J. Z. Werblowsky (Princeton: Princeton University Press, 1973).

40. David Biale, *Gershom Scholem: Kabbalah and Counter-History* (Cambridge: Harvard University Press, 1979), p. 155.

41. Bloom, "The Breaking of Form," *Deconstruction and Criticism*, ed. Bloom et al. (New York: Seabury Press, 1979), p. 4.

42. See, for example, Scholem, *Major Trends;* "The Meaning of the Torah in Jewish Mysticism," in *On the Kabbalah*, pp. 32–86; "The Name of God in the Linguistic Theory of the Kabbalah," *Diogenes* 79(1972): 59–80; and vol. 80 (1972): 164–94; and Biale's discussion in *Gershom Scholem*, pp. 79–122.

43. Gershom Scholem, "Revelation and Tradition as Religious Categories in Judaism," in *The Messianic Idea in Judaism*, p. 294.

44. Ibid., p. 295.

45. Derrida, "Ellipsis," *Writing and Difference*, pp. 294, 295.

46. Geoffrey Hartman, "Epiphony in Echoland," in *Saving the Text*, p. 48.

47. Letter to Franz Werfel, probably not mailed. Cf. the letters to Max Brod and Franz Werfel of December 1922 in *Briefe 1902–24*, ed. Max Brod (New York: Schocken, 1958), pp. 432 ff., cited in Marthe Robert, *From Oedipus to Moses: Freud's Jewish Identity*, trans. Ralph Mannheim (New York: Doubleday-Anchor, 1976), p. 173, n. 12.

48. Robert, *Oedipus to Moses*, p. 8.

49. Derrida, "Edmond Jabès," *Writing and Difference*, p. 74. Succeeding references to this essay are cited parenthetically in the text.

50. Franz Kafka, quoted in Hartman, *Criticism in the Wilderness*, p. 55.

Part Three

***

The Politics of
Displacement

Introduction:
Literariness—The Politics of Displacement

At every point the literary language oversteps the boundaries that literature apparently marks off; we need only consider the influence of *salons*,
the court, and national academies.

—F. DE SAUSSURE, *Course in General Linguistics*

IN RECENT YEARS BOTH MARXIST and deconstructive theorists have strongly
challenged the customary practice of attributing to literature some essential, autonomous property called "literariness." Both discourses find this
post-Kantian, aestheticizing conception to be fraught with certain political
implications; each, however, tends to assess these implications in a manner radically distinct from (though not simply opposed to) that of the
other.

While it is true that Herbert Marcuse considered "the aesthetic dimension" as a universal reservoir of potentially contestatory values, most contemporary Marxist theorists identify this domain as itself the product of
bourgeois culture and hence as an ideological rather than essential construct. Raymond Williams, for example, in his book *Marxism and Literature*, argues that the "aesthetic" and the "literary" did not attain their
present significance until the beginning of the nineteenth century, and
thus command only a relative rather than a universal bearing within the
confines of post-Romantic culture. Accordingly, it is only in a restricted
sense—and not as an ensemble of timeless textual properties—that the
term "literature" may appropriately be used today. "Literature" now des-
ignates

> a *particular, historically determined* form of writing, defined by the major
> forms of bourgeois society, instead of, as is customarily the case, a set of
> universal attributes which *all* major forms of writing, from Homer to
> Kafka, are held to have in common. [Tony Bennett, *Formalism and Marxism*,
> pp. 14–15]

The origins of this essentializing practice have been traced by Frank
Lentricchia directly to Kant, whose very attempt to differentiate the
"properly aesthetic" from the moral and the cognitive is seen as betraying

an ultimately ideological desire to deny the historical condition of all experience:

> [Kant's] intention of isolating the distinctive character of the aesthetic experience was admirable, but his analysis resulted in mere isolation. By barring that experience from the truth of the phenomenal world, while allowing art's fictional world entertainment value, he became the philosophical father of an enervating aestheticism which ultimately subverts what it would celebrate. [*After the New Criticism*, p. 43]

Kant's very distinction between the aesthetic and the quotidian thus is vitiated by the fact that the former exists only as a historical function, that "there is no such 'thing' as *literature*, no body of written texts which self-evidently bear on their surface some immediately perceivable and indisputable literary essence" (Bennett, p. 9). Condemning as an inherently ideological gesture—as a naturalization of class-specific practices—any such recourse to essentializing categories, these Marxist theorists all rely on a notion of ultimate historical foundation by which to displace the concept of the literary from a universal to a political register.

While for Derrida, too, there is "no such 'thing' as *literature*," it must be acknowledged at once that "literature" plays a role in his writings that, undermining the univocal status of *all* existing "things" (itself included), is finally irreducible to the Marxist relativization of the term. In *Dissemination*, for example, Derrida concurs with the Marxists' view that there are no ascribable essences that make literature literature; he similarly criticizes any conceptual category that aspires to universality, which "seems to aim toward the filling of a lack (a hole) in a whole that should not itself in its essence be missing (to) itself" (p. 56). If, however, Derrida agrees that the specificity of literature consists only in its lack of any proper self-identity—if the relationship between the literary and the nonliterary is thus compromised from the outset—he is quick to draw the unprecedented conclusion that literature must therefore function as *the exception to every thing else* as well: "at once the exception in the whole, the want-of-wholeness in the whole, and the exception to everything, that which exists by itself, alone, with nothing else, in exception to all." If, in other words, literature voids itself continually in its total absence of propriety—in its structural inability to constitute itself *as* a discrete entity—then everything that defines itself in opposition to the literary (which, in the Kantian schema, *is* "everything") necessarily will become infected by contact with this radical lack of self-sameness. Retaining the name "literature" (always used in quotation marks) as that which "breaks away from [the essentialist concept of] literature—away from what has always been conceived and signified under that name" (*Dissemination*, p. 3), Derrida would demonstrate that Kant's founding distinction inevitably must undo itself—not, as the Marxists claim, in its failure to recognize the historicity of the categories involved, but as a consequence of the subversive impact of "literature" on the integrity of these categories themselves. If, then, it is Marxism's goal to displace the literary from an essentialist into an ideological dimension ("displacement's politics"), it is Derrida's to indicate that the assumed stability of this very notion of the political—as a historically-determined, univocal "last instance"—is itself susceptible to displacement through the disseminative powers of "literature" ("politics' displacement").

The double genitive "displacement of politics" ("displacement's politics"/"politics' displacement") thus stands as an appropriate figure for the essays that follow, for each situates itself explicitly—albeit with differing emphases—within this complex space between literature and "literature," between Marxism and deconstruction. Paul de Man's treatment of Hegel's writings on the sublime, Michael Ryan's analysis of the metaphorics of liberalism, and Gayatri Spivak's reading of the question of "woman" in Derrida all bear the traces of an intricate process of negotiation between these critical discourses, all demonstrate how politics can be at once both the subject and the object of displacement.

While the term "displacement" is not fully elaborated upon in Paul de Man's "Hegel on the Sublime," the effects of this figure are very much in evidence in this major reappraisal of the relationship between aesthetics and politics in the Hegelian corpus. Although the notion of the sublime in late eighteenth- and early nineteenth-century thought has recently received extensive critical attention, de Man's interest in Hegel's use of the term concerns its unique function in the *Aesthetics* as the putative link between discourse and action, religion and law, language and politics. Finding the aesthetic moment to be marked in Hegel by "the conscious forgetting of a consciousness by means of a materially actualized system of notation or inscription," de Man explores the implications of this "forgetting" in Hegel's two irreconcilable accounts of the sublime, as both representation and trope. Whereas this first, mimetic model of the sublime—however negatively portrayed by Hegel—remains implicitly available for dialectical recuperation, the latter apostrophic model introduces a breach into the coherence of the narrative that precludes the possibility of *any* ultimate synthesis. Such a breach necessarily complicates Hegel's intended "passage," according to de Man, "from the aesthetic theory of the sublime to the political world of the law" insofar as this passage must assume the uncanny form of "a new linguistic model" no longer symbolic but "closer to that of the sign or trope, yet distinct from both in a way that allows for a concatenation of semiotic and tropological features." It is this perverse conflation of linguistic properties that allows de Man to read

Hegel's "prosaic" conclusion as "a stutter, or a broken record, [that] makes what it keeps repeating worthless and meaningless"—a conclusion which becomes "politically legitimate and effective" to the extent that Hegel's rhetoric ultimately functions as "the undoer of usurped authority" (including, by implication, the authority of the *Aesthetics* as well). Thus the relationship between aesthetics and politics in the *Aesthetics* is rendered simultaneously more direct and more oblique than in its customary interpretation: more direct, in that any consideration of the political now must pass through a preliminary consideration of the aesthetic; more oblique, in that the unresolvable complications of the aesthetic call into question the very assurance of any such "passage." It is this latter phenomenon that we may identify as the "displacement of politics" in de Man's reading of Hegel, for the political now paradoxically transforms itself into that which cannot constitute itself as a univocal politics.

A substantial disagreement concerning the "proper" locus of the political separates "Hegel on the Sublime" from Michael Ryan's critique of liberalism, "Deconstructive Philosophy and Social Theory." For if de Man addresses those (like Ryan) who question deconstruction's preoccupation with the literary, Ryan "responds" by urging a radical extension of deconstructive practice to extraliterary domains. Thus, where de Man locates the impasse of politics within the aesthetic, Ryan would displace this paradigm in tracing the effects of "literariness" within the body politic.

Quoting Rousseau's contention in *The Social Contract* that "the state if it is to have strength must give itself a solid foundation," Ryan argues that the modern capitalist state has founded itself only on a liberal ideology that actively represses its origin as figure rather than as stable ground. Although liberalism traditionally assumes that its twin doctrines of individual rights and state sovereignty actually furnish such a "solid foundation," Ryan demonstrates convincingly that "liberal institutions are not founded on anything that can be called 'real'"—hence their dependence on *force* to coerce what they cannot "naturally" command. "Liberalism can assert the doctrine of rights only as an abstract formal principle, because in substance liberal society, as capitalist society, must deny the right to property to workers and to women if it is to function rationally and scientifically." The notion of the sanctity of private property thus serves to legitimate a manifestly unequal distribution of wealth: while all, in theory, are entitled to property, the actual deeds of entitlement remain in the hands of a privileged male elite.

Yet inherent in this very concept of "natural" rights is an aporia that—parallel to Marx's predictions concerning the self-transcendence of capitalism—can be viewed as the lever by which liberalism will generate its own transformation. This aporia is located in the structure of property as such, in its essential lack of any determinate essence, of any "proper" self-identity: "Natural rights are never fully natural; they are displaceable, that is, contractable and socially exchangeable, from the outset. The anchor of liberalism—natural law—is thus made unstable by precisely that which the anchor supports or to which it gives rise—displacement." The capacity to be displaced, or transferred, is therefore less an accidental or secondary feature of property than its very condition of possibility—a condition which renders *im*possible any notion of stable self-identity. For if property is born as displacement, then it remains fundamentally alienable in deriving from a process which resists coming to rest at any point of self-sameness. Property, for Ryan, thus is infected with "literariness" from the start, is structurally incapable of serving the requirements of essential integrity on which liberalism depends—"and that," he rather optimistically concludes, "would mean the disappearance of property as liberalism conceives it."

If de Man's and Ryan's essays are joined by an implicit agon, Gayatri Spivak's essay "Displacement and the Discourse of Woman" identifies explicitly the source of its productive ambivalence: Derrida's frequent use of the figure of woman as an image for "literariness." Wondering overtly whether it is indeed possible for a woman—historically denied the status of subject—to write such essays at all, Spivak asks that her contribution be read as a record of struggle replete with revisions, temporal discontinuities, (self-)criticisms, second and even third thoughts; it is an analysis of displacement that earns something of an exemplary status in displacing *itself* continually.

"Displacement and the Discourse of Women" falls into two asymmetrical though interwoven parts, the first a critique of Derrida's treatment of "the woman question," the other an attempt to define a political feminist practice that conserves the best deconstruction has to offer. Spivak makes it clear why these two separate gestures are necessary: Derrida does not offer (himself as) a paradigm to be emulated by the feminist critic; certainly a fellow-traveler, he is, nevertheless, no "role-model." Challenging forcefully the norms of patriarchal discourse by reversing and reinscribing the traditionally subservient role of woman, deconstruction still requires, however, a *doubly*-displaced or "literary" image of woman as a pretext for its act of intervention. As Spivak undertakes brief readings of the sexual metaphorics of Derrida's texts, she finally concludes that "the woman who is the 'model' for deconstructive discourse remains a woman generalized and defined in terms of the faked orgasm and other varieties of denial." Hardly a role-model, indeed.

The feminist deconstructionist thus must "learn from Derrida's critique of phallocentrism and go somewhere else with it." Where Spivak goes is to "a certain historical 'differential' " which enables her to reread the texts of feminism with "a new look," to "rewrite the *social* text" of motherhood. While careful to avoid "the prospective historical or psychological continuity [that] is the idealist subtext of the patriarchal project," she learns to ask not "what is man?" but "what is man that the itinerary of his desire creates such a text?" Assuming for the first time, then, the privilege of serving as a questioning subject, she supplements such questioning with "the collective and substantive work of 'restoring' woman's history and literature."

All three of these essays therefore cross and recross the boundaries between "literature" and politics, history and rhetoric; though the route taken by each is distinctive, the terrain covered is common to all. What they demonstrate collectively is that literary theory need not "die" (as some have argued) once its object—literariness—has lost its privileged status; rather, the task of theory is just beginning in reconceptualizing the history of its past and in projecting the history of its future. Political considerations will remain central to both goals, even though our customary understanding of politics may itself become displaced in the process— for if, to paraphrase Saussure, "literature" oversteps the (political) boundaries that literature apparently marks off, then our project does not remain "merely" theoretical but necessarily affects the borders of our "national academies" as well.

—ANDREW PARKER

HEGEL ON THE SUBLIME

***

Paul de Man

JUST AS THE PLACE OF AESTHETICS in the canon of Hegel's works and in the history of its reception remains hard to interpret, the place of the sublime within the more restricted corpus of the *Aesthetics* itself is equally problematic. The fact that the same observation, with proper qualifications, applies to Kant as well, compounds the difficulty. The ensuing uncertainties help to account for the numberless confusions and misguided conflicts that clutter the stage of contemporary theoretical discourse on or around literature. One striking example of such a confusion is the principle of exclusion that is assumed to operate between aesthetic theory and epistemological speculation or, in a symmetrical pattern, between a concern with aesthetics and a concern with political issues.

The confusion has curious consequences. Derrida's name, for instance—to take an example that is both timely and closely familiar—is anathema to a number of literary academics, not so much because of his declared political opinions or positions, but because he has arrived at these positions by way of a professional philosopher's skills and interests; the same people who consider his influence nefarious can be quite tolerant toward writers or critics who are more flamboyantly radical in politics but who stay away from the technical vocabulary of philosophical cognition. On the other hand, Derrida is treated with a great deal of suspicion, if not downright hostility, by political activists, Marxist or other, for no other reason than that the canon on which he works remains "confined to philosophical and literary texts" and, hence, "confined to concepts and to language rather than to social institutions." Reactionaries deny him access to the aesthetic because he is too much of a philosopher, while proponents of political activism deny him access to the political because he is too concerned with questions of aesthetics. In both cases the aesthetic functions as the principle of exclusion: aesthetic judgment, or the lack of it, excludes the philosopher from access to literature, and the same aesthetic judgment, or the excess of it, excludes him from the political world. These symmetrical gestures, even if one does not approve of them, appear commonplace and easy to understand. Yet intellectual history, let alone actual philosophy, tells a very different story.

In the history of aesthetic theory since Kant, aesthetics, far from being a principle of exclusion, functions as a necessary, though problematic, articulation. In Kant, the articulation of the First Critique with the Third, of the schemata of theoretical reason with those of practical reason, has to occur by way of the aesthetic, successful or not. Aesthetic theory is critical philosophy to the second degree, the critique of the critiques. It critically examines the possibility and the modalities of political discourse and political action, the inescapable burden of any linkage between discourse and action. The treatment of the aesthetic in Kant is certainly far from conclusive, but one thing is clear: it is epistemological as well as political through and through. That several intellectual historians, American as well as European, have been able to claim the reverse and to assert that the aesthetic in Kant is "free from cognitive and ethical consequences," is their problem, not Kant's.

Hegel, on the same question, is even more explicit. The link between politics, art, and philosophy, by way of a philosophy of art or aesthetics, is built into his system, not in the unreflected sense that aesthetics is concerned with the political as its subject matter, but in the much further-reaching sense that, here again, the trajectory from political to intellectual reality, the passage, in Hegel's terminology, from the objective to the absolute spirit, passes by necessity through art and through the aesthetic as critical reflection on art. At this crucial junction between the most advanced stages of political thought, in the attempt to conceive of the state as historical act, and philosophical thought, Hegel, in the *Encyclopedia of the Philosophical Sciences*, situates art. How this is to be understood is certainly not a simple matter, neither in itself nor in the history of Hegel reception as it has come down to us and as Hegel is read today; it depends on the reading of Hegel's own treatment of the aesthetic in the late *Lectures on Aesthetics*. But one thing can be ascertained from the start: by dint of the structure of the Hegelian system, the consideration of aesthetics only makes sense in the context of the larger question of the relationship between the order of the political and the order of philosophy. This would imply that since in Hegel the aesthetic belongs to a more advanced but proximate stage of speculative thought than political reflection, truly productive political thought is accessible only by way of critical aesthetic theory. The last thing this sentence means, in Hegelian or in any other terms, is that political wisdom belongs to what we ordinarily call aestheticism. What it might mean, to return to our initial example, is that someone like Derrida is politically effective because of, and not in spite of, his concentration on literary texts. This would be borne out by the historical fact that some of the most incisive contributions to political thought and political action have come from "aesthetic" thinkers. Marx himself, whose

German Ideology is a model of critical procedure along the lines of Kant's Third Critique, is a case in point—as are, closer to our own times, the writings of Walter Benjamin, Lukács, Althusser, and Adorno. But the work of what are then called "aesthetic thinkers" bears little resemblance to what nineteenth- and twentieth-century literary history identifies as aestheticism, for the work of these thinkers precludes, for example, any valorization of aesthetic categories at the expense of intellectual rigor or political action, or any claim for the autonomy of aesthetic experience as a self-enclosed, self-reflexive totality.

These preliminary remarks lead up to the task of interpreting Hegel's *Lectures on Aesthetics*. Despite the considerable amount of philosophical and critical talent that has been mobilized in its behalf, this task has proved to be very difficult. In no instance has it been possible to reach a consistent reading, especially when, as is the case for Heidegger and for Adorno, the reading of the *Aesthetics* has to become part of a general critical reading of Hegel himself by way of such key concepts as that of *Aufhebung* or of the dialectic itself. For, at first sight, the *Aesthetics* appear as the most blandly orthodox and dogmatic of the late writings, at a moment when the magisterial exposition of the system seems to have reached a stage of deadly mechanical didacticism. One either has to reduce the *Aesthetics*, as Heidegger does, to the gnomic wisdom of its most enigmatic pronouncements—the end of art, the sensory appearance of the idea—and treat it as the mute sphinx that ends all conversation, or like Adorno and some of his followers, one has to consider it as the Achilles heel of an entire system, in the very specific sense that the *Aesthetics* would be the place where the inadequacy of Hegel's theory of language would be revealed. The dispersal over the entirety of the collected works of an implicit conception of language that is never formulated weighs heavily on the enterprise and privileges passages and texts in which such a theory, by coming closest to being stated, would finally reveal its shortcomings. Peter Szondi, a literary historian who was himself close to Adorno, locates such a place in the *Aesthetics*, in Hegel's discussion of allegory and metaphor: "Hegel's often derogatory characterization of these poetic resources (allegory and metaphor) allows us to see why they lay almost entirely beyond his understanding. In asking for these reasons, the limits of Hegel's aesthetics also become visible . . . It is Hegel's inadequate conception of the nature of language that is to be blamed for his failure."[1] The importance of the *Aesthetics* as a possible point of entry into a critique of the dialectic stands clearly revealed in this quotation. For if the inadequacy of the *Aesthetics* is due to an inadequate theory of language, then this is bound to contaminate the logic, the phenomenology of cognition, and ultimately all the essential claims of the system. It is the considerable merit of this approach to have directed attention to what are indeed the determining issues and even the determining passages involved in the interpretation of the *Aesthetics*.

A reading thus sensitized to linguistic terminology and to the problematics of language can hope to displace the received ideas, oracular or dogmatic, by which the interpretation of the *Aesthetics* had been brought to a standstill. It also allows for an extension of the textual corpus from the *Aesthetics* to considerations on language that occur elsewhere in Hegel's works, in the *Encyclopedia*, the *Science of Logic*, or the *Phenomenology of the Spirit*. It allows for a linkage between the theory of language, of the subject, and of sensory perception. Finally, it should clarify the relationship of art and literature to the dimension of pastness that is a necessary component of any discourse involving history. If art, the sensory or (better) the phenomenal manifestation of the idea, belongs for us, as the *Aesthetics* assert, to the past, then this pastness is to be a function of its phenomenality, of its mode of appearance. Where and how is it then, in the system of Hegel's writings as a whole, that the idea appears, and why does this specifically aesthetic moment belong necessarily to the past?

The *Aesthetics* seem to provide only banal and empirical answers to these questions. In the wake of Winckelmann and Schiller, it historicizes the problem in the ideologically loaded genealogy of the modern as derived from the classical, Hellenic past, thus creating an illusion of misplaced concreteness that is responsible for a great deal of poor historiography from the early nineteenth century to the present. These historical fallacies run parallel to a concept of language in which the all-important distinction between the symbolic and the semiotic aspects of language is eroded. Since this happens most distinctively in the *Aesthetics*, one sees the need for a detour out of this work into other Hegel texts in which the discussion of the same issue is less blurred by romantic ideology.

This allows for a more precise answer to the question about the appearance of the idea. Most clearly in the *Encyclopedia*, but in the *Logic* as well, the idea makes its appearance on the mental stage of human intelligence at the precise moment when our consciousness of the world, which faculties such as perception or imagination have interiorized by way of recollection *(Erinnerung)*, is no longer experienced but remains accessible only to memorization *(Gedächtnis)*. At that moment, and at no other, can it be said that the idea leaves a material trace, accessible to the senses, upon the world. We can perceive the most fleeting and imagine the wildest things without any change occurring to the surface of the world, but from the moment we memorize, we cannot do without such a trace, be it as a knot in our handkerchief, a shopping list, a table of multiplication, a psalmodized sing-song or plain chant, or any other memorandum. Once such a *notation* has occurred, the inside-outside metaphor of experience and signification can be forgotten, which is the necessary (if not sufficient) condition for thought *(Denken)* to begin. The aesthetic moment in Hegel occurs as the conscious forgetting of a consciousness by means of a materially actualized system of notation or inscription.

This conclusion is derived from a section of the *Encyclopedia* entitled "Psychology" that follows upon the section entitled "Consciousness." Nothing even remotely similar seems to be stated in the public theses or arguments of the *Aesthetics*. Lest we assume that the later, professorial Hegel was so busy thinking that he "forgot" his former more speculative self—a supposition that flatters the source of our livelihood way beyond its deserts—such inconsistencies in a systematic philosopher are unlikely; they tend to memorize their own writings very thoroughly. Much more likely is the assumption that similar or equivalent assertions actually occur in the *Aesthetics* but that, for a variety of reasons, the passages in which they occur have been overlooked, misunderstood, or censored. If Hegel's theory of memorization has any merit at all, the pressure of its power would indeed have to make itself felt in the *Aesthetics*, in however oblique or disguised a way. Chapter II of the section on symbolic art entitled "Die Symbolik der Erhabenheit" (The Symbolics of the Sublime),[2] a chapter that immediately precedes the section on comparative art forms singled out by Peter Szondi as the nether point of Hegel's aesthetic sensibilities, is one of the places in the *Aesthetics* where it surfaces.

The first hint one gets of this is in Hegel's rather shabby treatment of Kant early on in the chapter. We are told that Kant's treatment of the sublime is long-winded but still of some interest—and the reasons given for this interest are altogether cogent. But by putting so much emphasis on the particularity of the affects in which he rightly chose to locate it, Kant has trivialized the sublime. It is open to question whether Hegel does justice here to Kant's concept of affect *(Gemüt)*, but one can surmise the reasons for his impatience with Kant's interest in affect and mood. For if the aesthetic, in Hegel, is indeed akin, in some way or another, to memorization, then it has little concern for particularized emotions, and any self-conscious sentimentalization had better be checked from the start.

More revealing, perhaps, though still merely formal, is the place Hegel allots to the sublime in the dialectical continuum of the various art forms. "We find the sublime first and in its original form primarily in the Hebraic state of mind and in the sacred texts of the Jews" (p. 480). The association of the sublime with the poetry of the Old Testament is a commonplace, especially in Germany after Herder, but Hegel's reasons are of interest. Hebraic poetry is sublime because it is iconoclastic; it rejects art as plastic or architectural representation, be it as temple or as statue. "Since it is impossible to conceive of the image of the divine that would in any degree be adequate, there is no place for the plastic arts in the sublime sacred art of the Jews. Only the poetry of a representation that manifests itself by means of the *word* will be acceptable." In its explicit separation from anything that could be perceived or imagined, the word indeed appears here as the inscription which, according to the *Encyclopedia*, is the first and only phenomenal manifestation of the idea. Monuments and statues made of stone and metal are only pre-aesthetic. They are sensory appearances all right, but not, or not yet, appearances of *the idea*. The idea appears only as written inscription. Only the written word can be sublime, to the precise extent that the written word is neither representational, like a perception, nor imaginative, like a phantasm.

The section on the sublime confirms this formal affirmation and develops some of its implications and consequences. In the process, it soon becomes apparent that the sublime in Hegel differs considerably from the post-Longinian sublime of those of his predecessors, to borrow a suggestive listing from Meyer Abrams's very useful chapter on the sublime in *The Mirror and the Lamp*, such as John Dennis, Bishop Lowth, and Herder,[3] a tradition which has survived in the American interpretation of Romanticism in Wimsatt, Abrams, Bloom, Hartman, and Weiskel; it was finally ironized, though not necessarily exorcised, in Neil Hertz's remarkable essay "Lecture de Longin"[4]—which remains conveniently hidden from the tradition by appearing, of all places, in Paris, where no one can appreciate what is at stake in this closely familial romance. The most conspicuous, though not the most decisive of these differences, resides in the disappearance of the familiar oppositions between poetry and prose, or between the sublime and the beautiful. The sublime for Hegel *is* the absolutely beautiful. Yet nothing sounds less sublime, in our current use of the term, than the sublime in Hegel. That it marks an open break with the linguistic model of the symbol that pervades all sections of the *Aesthetics* is visible from the start; already in the introductory section on Kant it is said that in the sublime "the actual *symbolic* character" of the work of art vanishes (p. 468). What this involves however becomes clear only as the inner logic of the passage is allowed to unfold.

The moment Hegel calls sublime is the moment of radical and definitive separation between the order of discourse and the order of the sacred. The necessity to isolate such a moment is forced upon him by the concept of language as symbol to which the *Aesthetics* is firmly committed—without which, indeed, no such topic as the aesthetic could come into being. The phenomenality of the linguistic sign can, by an infinite variety of devices or turns, be aligned with the phenomenality, as knowledge (meaning) or sensory experience, of the signified toward which it is directed. It is the phenomenalization of the sign that constitutes signification, regardless of whether it occurs by way of conventional or by way of natural means. The term phenomenality here implies not more and not less than that the process of signification, in and by itself, can be known, just as the laws of nature as well as those of convention can be made accessible to some form of knowledge.

The constraint to abandon this claim arises, in Hegel, from the classical and, in this case, Kantian critical process to discriminate between modes of cognition and to separate the knowledge of the natural world from the knowledge of how knowledge is achieved, the separation between mathematics and epistemology. In the history of art, it corresponds to the moment when the infinite diffusion and dispersal of what Hegel calls the "single substance" *(die eine Substanz)* that stands beyond the antinomy of light and the shapeless, singularizes itself in the designation of this absolute generality as the sacred or as god. It is the passage from pantheistic to monotheistic art, the passage, in Hegel's picture-book but by no means innocent history, from Indian to Mohammedan poetry. The relationship between pantheism and monotheism in the history of art and religion (since, up to this point, it would be impossible to distinguish between them) is like the relationship between natural science and epistemology: the concept of mind (be it as Locke's understanding, Kant's *Vernunft*, or Hegel's Spirit) is the monotheistic principle of philosophy as the single field of unified knowledge. The monotheistic moment (which in Hegel is not or not yet the sublime) is essentially verbal and coincides with the fantastic notion that *die eine Substanz* could be given a name—such as, for instance, *die eine Substanz*, or the One, or Being, or Allah, or Jahwe, or I— and that this name could then function symbolically, yielding knowledge and discourse. From this moment on, language is the deictic system of predication and determination in which we dwell more or less poetically on this earth. In conformity with his tradition and with his place in the ongoing discourse of philosophy, Hegel understands this moment as a relationship between mind and nature constituted by negation. But behind this familiar and historically intelligible dialectical model stands a different reality. For it is one thing to assert that absolute knowledge accomplishes its labor by way of negation, another thing entirely to assert the possibility of negating the absolute by allowing it, as in this passage, to enter in an unmediated relationship with its other. If "mind" and "nature" stand, in fact, for the absolute and its other, then Hegel's narrative resembles that of dialectical sublation or upheaval *(Aufhebung)* only on a first level of understanding.

The difficulty of the passage—the section entitled "Die Kunst der Erhabenheit" and the subsequent chapter, "The conscious symbolism of comparative art forms," (pp. 479–539)—stems indeed from the interference of a dialectical with another, not necessarily compatible, pattern of narration. When we read of a hidden god who has "withdrawn into himself and thus asserted his autonomy against the finite world, as pure interiority and substantive power," or hear that in the sublime, the divine substance "becomes truly manifest" (p. 479) against the weakness and the ephemerality of its creatures, then we easily understand the pathos of this servitude as praise of divine power. The language of negativity is then a dialectical and recuperative moment, akin to similar turns that Neil Hertz has located in Longinus' treatise. Hegel's sublime may stress the distance between the human discourse of the poets and the voice of the sacred even further than Longinus, but as long as this distance remains, as he puts it, a *relationship* (pp. 478, 481), however negative, the fundamental analogy between poetic and divine creation is preserved. Yet the narrative generated on the level of the dialectic does not correspond to the implicit narrative of the section. To be *erhaben* (sublime) is not the same, it appears, as to be *erhoben* or *aufgehoben*, however close the two words may be in sound and despite Hegel's occasional substitution of one term for the other (as at the bottom of p. 483). If one considers the one example of sublimity that Hegel and Longinus share in common, the *fiat lux* from Genesis, then the complexity of the passage begins to appear (pp. 481, 484). Hegel quotes "And God said, let there be light and there was light" to illustrate that the relationship between God and man is no longer natural or genetic, and that God cannot be considered as a progenitor. For *zeugen* (to engender) Hegel wishes to substitute *schaffen* (to create), but creation then has a stronger negative connotation than the term normally implies; no polarity of young/old or male/female should suggest a familial hierarchy. The hierarchy is much starker. Creation is purely verbal, the imperative, pointing, and positing power of the word. The word speaks and the world is the transitive object of its utterance, but this implies that what is thus spoken, and which includes us, is not the subject of its speech act. Our obedience to the word is mute: "the word . . . whose command to be also and actually posits what is without mediation and in *mute* obedience" (p. 480, italics mine). If the word is said to speak through us, then we speak only as a ventriloquist's dummy, also and especially when we pretend to talk back. If we say that language speaks, that the grammatical subject of a proposition is language rather than a self, we are not fallaciously anthropomorphizing language but rigorously grammatizing the self. The self is deprived of any locutionary power; to all intents and purposes it may as well be mute.

Yet *das Daseiende* that language produces speaks, and even writes, a great deal in Hegel, and in an interesting variety of ways. First of all, it quotes. Scripture quotes Moses who quotes God and it makes use, in Genesis, of the fundamental rhetorical modes of representation: mimesis, in Plato's sense, as reported speech *(erlebte Rede)*, as in the sentence: "And God said 'Let there be light,'" as well as, closely intertwined with the former, diegesis or indirect speech *(erzählte Rede)*, as in the sentence: "And God called the light day. . . ." At this level, the distinction between the two modes of locution is not important, since mimesis and diegesis are part of the same system of representation; the mimesis can always be considered as encased in a third-person narrative (and he said ". . ."). But none of these utterances are mute in the sense of being merely passive or devoid of reflexive knowledge. Quotations can have considerable performative power; indeed, a case could be made that only quotations have such power. Even hidden quotations are not mute: a plagiarizer who gets caught may be dumb but he (or she) is no dummy. They are, however, devoid of positional power: to quote the marriage vows allows one to perform a marriage but not to posit marriage as an institution. And quotations certainly carry a considerable cognitive weight: if, as Longinus implies, the sublime poet here is Moses himself, then the question of the veracity of Moses' testimonial is bound to arise, that is to say, a cognitive critical inquiry is inevitably linked to the assertion of linguistic positional force. This accounts for the fact that in a statement such as "Let there be. . . ," *light* is indeed the privileged object of predication, rather than life (Let there be life) or humanity (Let there be woman and man). "Light" names the necessary phenomenality of any positing *(setzen)*. The convergence of discourse and the sacred which, in the choice of example and in Hegel's commentary on it, is not in question, occurs by way of phenomenal cognition. No matter how strongly the autonomy of language is denied, as long as the language can declare and know its own weakness and call itself mute, we remain in a Longinian mode. Pascal's paradox applies: "In a word, man knows that he is miserable. Thus, he is miserable since that is what he is. But he is very great inasmuch as he knows it." A dialectized sublime is still, as in Longinus, an intimation of poetic grandeur and immortality.

A little further in the text, Hegel mentions another example of the sublime also taken from Scripture, this time from the Psalms. In this case, the rhetorical mode is not a mimesis-diegesis system of representation but direct apostrophe. And what is said in apostrophe is curiously different from what is shown and told in representation although, or rather, because it also has to do with light: "Light is your garment, that you wear; you stretch out the heavens like a curtain . . ."[5] The juxtaposition of the two quotations, marked, however discreetly, by the symmetrical position of God and man ("von Seiten Gottes her" [p. 481] and "von Seiten des Menschen" [p. 484]), is quite amazing. The garment is a surface *(ein äusseres Gewand)*, an outside that conceals an inside. One can understand this, as Hegel does, as a statement about the insignificance of the sensory world as compared to the spirit. Unlike the *logos*, it does not have the power to posit anything; its power, or only discourse, is the knowledge of its weakness. But since this same spirit also, without mediation, *is* the light (p. 481), the combination of the two quotations states that the spirit posits itself as that which is unable to posit, and this declaration is either meaningless or duplicitous. One can pretend to be weak when one is strong, but the power to pretend is decisive proof of one's strength. One can know oneself, as man does, as that which is unable to know, but by moving from knowledge to position, all is changed. Position is all of a piece, and moreover, unlike thought, it actually occurs. It becomes impossible to find a common ground for or between the two quotations, "Let there be light" and "Light is your garment." To pretend, as Hegel does, that the first, which could be called Longinian, corresponds to the sublime as seen from the perspective, or the side, of God *(von Seiten Gottes)*, whereas the second corresponds to the sublime as seen from the perspective of man, does not, of course, mitigate or suspend the incompatibility. Within the monotheistic realm of *die eine Substanz*, no such thing as a human perspective could exist independently of the divine, nor could one speak of a "side" of the gods (as one speaks of the "côté de chez Swann"), since the *parousia* of the sacred allows for no parts, contours, or geometry. The only thing the misleading metaphor of a two-sided world accomplishes is to radicalize the separation between sacred and human in a manner that no dialectic can surmount *(aufheben)*. Such is indeed the declared thesis of the chapter, but it can only be read if one dispels the pathos of negation that conceals its actual force.

It is not insignificant that the reading only emerges from a combination of two rhetorical modes, that of representation and that of apostrophe. Paradoxically, the assumption of praise, in the Psalms, undoes the ground for praise established in Genesis. Apostrophe is the mode of praise par excellence, the figure of the ode. The strength of Hegel's choice of example makes clear that what the ode praises is not what it addresses ("la prise de Namur," Psyche or God)—for the light that allows the addressed entity to appear is always a veil—but that it always praises the veil, the device of apostrophe as it allows for the illusion of address. Since the ode, unlike the epic (which belongs to representation), knows exactly what it does, it does not praise at all, for no figure of speech is ever praiseworthy in itself. The passage reveals the inadequacy of the Longinian model of the sublime as representation. Apostrophe is not representation; it occurs independently of any report, be it as quotation or narration, and when it is put on a stage, it becomes ludicrous and cumbersome. Whereas representation can be shown to be a form of apostrophe, the reverse is not true. Apostrophe is a figure or a trope, as is clear from Hegel's next quotation from the Psalms, in which the garment becomes face: "Thou hidest thy face, they are troubled" (*Verbirgst du dein Angesicht, so erschrecken sie*— Psalm 106). The light from "Licht ist dein Kleid" is preserved in the German word *Angesicht*. The trope of face-giving is a particularly effective way by which to be drawn into the entire transformational system of tropes. When language functions as trope, and no longer only as representation, the limits of the Longinian sublime as well as of its considerable powers of recuperation, including the power of self-ironization,[6] are reached. As the section develops, the divergence between Hegel and Longinus becomes nearly as absolute as the divergence between man and God that Hegel calls sublime. Yet the two discourses remain intertwined as by a knot that cannot be unraveled. The heterogeneity of art and of the sacred, first introduced as a moment in an epistemological dialectic, is rooted in the linguistic structure in which the dialectic is itself inscribed.

In Hegel, too, there appears to be, at this point, a recuperative corollary to the declared otherness of the divine. It takes the form of a reassertion of human autonomy as ethical self-determination, "the judgment concerning good and evil; and the decision for the one over the other is now displaced in the subject itself" (p. 485). Out of this originates "a positive relationship to God" in the form of a legal system of reward and punishment. Before blaming or congratulating Hegel for this conservative individualism, one should try to understand what is involved in this passage from the aesthetic theory of the sublime to the political world of the law. Recuperation is an economic concept that allows for a mediated passage or crossing between negative or positive valorization: Pascal's pensée on human grandeur and misery that has just been quoted is a good example of how an absolute lack can be turned into an absolute surplus. But the definitive loss of the absolute experienced in the sublime puts an end to such an economy of value and replaces it with what one could call a critical economy: the law (*das Gesetz*) is always a law of differentiation (*Unterscheidung*), not the grounding of an authority but the unsettling of an authority that is shown to be illegitimate. The political in Hegel originates in the critical undoing of belief, the end of the current theodicy, the banishment of the defenders of faith from the affairs of the state, and the transformation of theology into the critical philosophy of right. The main monarch to be thus dethroned or de-sacralized is language, the matrix of all value systems in its claim to possess the absolute power of position. *Setzen* becomes *das Gesetz* as the critical power to undo the claim to power, not in the name of absolute or relative justice, but by its own namelessness, its own ordinariness. To pursue this would take us into the two treatises that have to be considered conjointly and in the wake of the *Aesthetics:* the *Principles of the Philosophy of Law* and the *Lectures on the Philosophy of Religion.* What matters for our topic is that the necessity to treat these two alien political forces, law and religion, is established in the *Aesthetics,* and specifically in the aesthetics of the sublime. That this would have been taken to state the reverse of what it states confirms the strength of Hegel's analysis; the same fate will soon befall similar assertions in Kierkegaard and Marx and, in our own times, in Walter Benjamin.

There can be no better preparation for a critical reading of the Philosophies of Law and Religion than the immediate sequel to the theory of the sublime in what Hegel somewhat cryptically calls "comparative" art forms. The separation, still suspended in deliberate ambivalence in the sublime, now carries out the law of its occurrence into its next stage. Language as symbol is replaced by a new linguistic model, closer to that of the sign and of trope, yet distinct from both in a way that allows for a concatenation of semiotic and tropological features. This complication is reflected in the curious combination of art forms that make up this section: some of them, like metaphor, allegory, and something called image (*Bild*) are more or less straightforward tropes, but others, such as fable, proverb, and parable, are minor literary genres that seem to be of an entirely different order. What is being undone, in each of these instances and cumulatively in their succession, is not the duel structure of signification as a combination of sign and meaning which the symbol is assumed to overcome. Rather it is the homology, in each of the particular genres of tropes, of the structure that defines them with the structure that defines the symbol. Hegel first makes the point by way of a traditional inside/outside polarity. Even the symbol does not simply coincide with the entity it symbolizes; it requires the mediation of an understanding to cross the borderline that keeps it "outside" this entity. The relationship between sign and meaning, however, in the symbol, is dialectical. But now "this exteriority, since it was already latently (*an sich*) available in the symbol, must also be posited" (*Diese Äusserlichkeit aber, da sie* an sich *im Symbolischen vorhanden ist, muss auch gesetzt werden*). This *Gesetz der Äusserlichkeit* implies that the principle of signification is now itself no longer animated by the tensions between its dual poles, but that it is reduced to the preordained motion of its own position. As such, it is no longer a sign-producing function (which is how Hegel valorized the sign in the *Encyclopedia*), but the quotation or repetition of a previously established semiosis. Neither is it a trope, for it cannot be closed off or replaced by the knowledge of its reduced condition. Like a stutter, or a broken record, it makes what it keeps repeating worthless and meaningless. The passage is itself the best illustration of this. Completely devoid of aura or *éclat*, it offers nothing to please anyone: it deeply distresses the aesthetic sensibilities of a symbolist like Peter Szondi, but it also spoils the fun of playful semioticians and reduces to nought the pretensions of the solemn ones, including the pathos of rhetorical analysis. Such passages are, of course, the ones to look out for in as pathos-laden a canon as Hegel's.

The spatial metaphor of exteriority *(Aüsserlichkeit)* is not adequate to describe the knowledge that follows from the experience of the sublime. The sublime, it turns out, is self-destroying in a manner without precedent at any of the other stages of the dialectic. "The difference between the present stage (that of the comparative art forms) and the sublime . . . is that the sublime relationship is completely eliminated *(vollständig fortfällt)*." There is nothing left to lift up or to uplift. This is the case with very few, if any, other key terms in Hegel's vocabulary, in the *Aesthetics* or elsewhere. One sees perhaps better what is implied by stating the process in temporal terms. This may also establish contact with the theory of symbol and sign, of subject and memorization, that was derived from the reading of section 20 of the *Encyclopedia*.

The alignment of literal and figural discourse in a figuration, by way (for example) of resemblance in the case of a metaphor, is now called by Hegel a *comparison*. The emphasis falls on the deliberate and conscious nature of this gesture. The juxtaposition of the two aspects of the figure is neither a genuine relationship nor a contractual convention, but an arbitrary positioning *(Nebeneinander-gestelltsein*, p. 487). The function of art is to make it appear as a discovery, when it is in fact preestablished by the one who claims to discover it. The illusion of discovery is consciously and cunningly contrived by means of a faculty Hegel calls *Witz* and which is very remote from natural genius in Kant or Schiller. Wit discovers nothing that is new or that was hidden; it invents only in the service of redundancy and reiteration. In temporal terms, it projects into the future what belongs to the past of its own invention and repeats as if it were a finding what it knew all along. This apparent reversal of past and present (metalepsis) is in fact no reversal at all, for the symmetrical equivalence of the sacrificed future is not an understood, but a trivialized past. Yet this bleak and disappointing moment, in all its sobriety, is also the moment in the *Aesthetics* when we come closest to the fundamental project of speculative philosophy. As we know from section 20 of the *Enclyclopedia*, the starting point of philosophy, "the simple expression of the existing subject as thinking subject" (I, p. 72), is equally arbitrary and pretends to verify its legitimacy in the sequential unfolding of its future until it reaches the point of self-recognition. Like the work of art, the subject of philosphy is a reconstruction a posteriori. Poets and philosophers share this lucidity about their enterprise.

Rather than putting it in terms that suggest deceit and duplicity, one could say that the poet, like the philosopher, must *forget* what he knows about his undertaking in order to accede to the discourse to which he is committed. Like all writers who happen to think wittily of some figure of language and then keep it embalmed, so to speak, in the coffin of their memory (or, in some cases, in an actual wooden box) until the day they will compose the text that proclaims to discover what they themselves had buried, poets know their figures only by rote and can use them only when they no longer remember or understand them. No actual bad faith is involved in such a process unless, of course, one claims transcendental merits for a move that pertains to the ethics of survival rather than of heroic conquest. At the end of the section in the *Aesthetics* on symbolic form, after the reversal of the sublime, writing is structured like memorization, or, in the terminology of Hegel's system, like thought. To read poets or philosophers thoughtfully, on the level of their thought rather than of one's or their desires, is to read them by rote. Every poem *(Gedicht)* is a *Lehrgedicht* (p. 541), whose knowledge is forgotten as it is read.

We can put this in still another way, providing a suitably arbitrary link with the political themes mentioned at the beginning of this paper. Hegel describes the inexorable progression from the rhetoric of the sublime to the rhetoric of figuration as a shrinkage from the categories of critical language that are able to encompass entire works, such as genre, to terms that designate only discontinuous segments of discourse, such as metaphor or any other trope. His own language becomes increasingly contemptuous of these sub-parts of the aesthetic monument. He calls them inferior genres *(untergenordnete Gattungen)*, only *(nur)* images or signs "deprived of spiritual energy, depth of insight, or of substance, devoid of poetry or philosophy." They are, in other words, thoroughly prosaic. They are so, however, not because of some initial shortcoming in the poet who uses these art forms rather than the major representational genres— epic, tragedy—but as the consequence of an inherent linguistic structure that is bound to manifest itself. In this entire development, there is no moment that could be reduced to avoidable accidents or contingencies, least of all the moment when poetic skills are shown to be themselves contingent and accidental. The infrastructures of language, such as grammar and tropes, account for the occurrence of the poetic superstructures, such as genres, as the devices needed for their oppression. The relentless drive of the dialectic, in the *Aesthetics*, reveals the essentially prosaic nature of art; to the extent that art is aesthetic, it is also prosaic—as learning by rote is prosaic compared to the depth of recollection, as Aesop is prosaic compared to Homer, or as Hegel's sublime is prosaic compared to Longinus'. The prosaic, however, should not be understood in terms of an opposition between poetry and prose. When the novel, as in Lukács's interpretation of nineteenth-century realism, is conceived as an offspring, however distant or elegiac, of the epic, then it is anything but prosaic in Hegel's sense. Nor would Baudelaire's prose versions of some of the poems of *Les fleurs du mal* be prosaic as compared to the metered and rhymed diction of the originals; all one could say is that they bring out the prosaic element that shaped the poems in the first place. Hegel summarizes his conception of the prosaic when he says: "It is in the slave that prose begins" (*Im Sklaven fängt die Prosa an*" [p. 497]). Hegel's *Aesthetics*, an essentially prosaic discourse on art, is a discourse of the slave because it is a discourse of the figure rather than of genre, of trope rather than of representation. As a result, it is also politically legitimate and effective as the undoer of usurped authority. The enslaved place and condition of the section on the sublime in the *Aesthetics*, and the enslaved place of the *Aesthetics* within the corpus of Hegel's complete works, are the symptoms of their strength. Poets, philosophers, and their readers lose their political impact only if they become, in turn, usurpers of mastery. One way of doing this is by avoiding, for whatever reason, the critical thrust of aesthetic judgment.

NOTES

1. Peter Szondi, *Poetik und Geschichtsphilosophie I* (Frankfurt am Main: Suhrkamp, 1974), pp. 390, 396; my translation.
2. Georg Wilhelm Friedrich Hegel, *Vorlesungen über die Aesthetik I*, vol. 13, *Werke* (Frankfurt am Main: Suhrkamp, 1970), pp. 466–546. All further references to this work are cited in the text.
3. M. H. Abrams, *The Mirror and the Lamp* (New York: Oxford University Press, 1953), pp. 72–78.
4. Neil Hertz, "Lecture de Longin," *Poétique* 15 (1973):292–306. An English translation of Hertz's essay has been published in the March 1983 issue of *Critical Inquiry*.
5. Hegel quotes from the Lutheran Bible. The King James Version is much less assertive: "Who coverest thyself with light as with a garment . . ." (Psalm 104).
6. Hertz, "Lecture," pp. 305–6.

DECONSTRUCTION AND SOCIAL THEORY
The Case of Liberalism

***

Michael Ryan

I AM INTERESTED IN USING deconstructive philosophy in social theory, especially in the critique of liberalism. Deconstructive philosophy criticizes some of the assumptions of "liberal Reason."[1] If, as Roberto Unger and Thomas Spragens suggest,[2] there is a relationship between liberal Reason and liberal politics, between how knowledge is conceived and described by liberal philosophers and how liberal political, juridical, social, and economic institutions are constituted, then it would seem to follow that a philosophy that offers a post-liberal description of Reason might also imply a post-liberal set of social institutions.

According to Spragens, liberal Reason consists of the following:

1. The assumptions and methods of the previously dominant Aristotelian-Scholastic tradition are mistaken and must be fundamentally revised or supplanted before genuine "natural philosophy" can be possible.
2. Human understanding, guided by the "natural light" of reason, can and should be autonomous. Moreover, it constitutes the norm and the means by reference to which all else is to be measured.
3. It is possible and necessary to begin the search for knowledge with a clean slate.
4. It is possible and necessary to base knowledge claims on a clear and distinct, indubitable, self-evident foundation.
5. This foundation is to be composed of simple, unambiguous ideas or perceptions.
6. The appropriate formal standards for all human knowledge are those of the mathematical modes of inquiry.
7. The key to the progress of human knowledge is the development and pursuit of explicit rules of method.
8. The entire body of valid human knowledge is a unity, both in method and in substance.
9. Therefore, human knowledge may be made almost wholly accessible to all men, provided only that they not be defective in their basic faculties.

10. Genuine knowledge is in some sense certain, "verifiable," and capable of being made wholly explicit.

11. Knowledge is power, and the increase of knowledge therefore holds the key to human progress.[3]

Deconstructive philosophy questions the normativization of the logos or the natural light of reason, and it doubts the possibility of a clean slate absolutely free of historical presuppositions. Also, it doubts the possibility of a self-evident foundation or of fully unambiguous ideas. It questions the absolute certainty of the mathematical model, the sanctity of formal methods abstracted from practical situations, and the unity of knowledge. And it questions, finally, the possibility of absolute certainty or determinacy. Much of this skepticism has to do with the nature and use of the language with which knowledge must necessarily be communicated.

I will be concerned with the way certain deconstructive arguments apply to the relationship between knowledge and politics in liberal theory. If liberal epistemology is subject to a deconstructive critique, does the argument carry over to liberal social institutions? To a certain extent, the question has already been answered in the affirmative, though in a non-deconstructive vocabulary, by Unger and Spragens. Unger locates irreducible antinomies within liberal theory, while Spragens suggests that liberalism produces technocratic and irrationalist consequences that can be avoided only by a thorough reconstruction of liberal Reason. This reconstruction would favor the emphasis on practice, indeterminacy, and difference advocated by Derrida.

Nevertheless, while Spragens and Unger offer sound criticisms of liberalism, they do not offer a significantly new conceptual or epistemological rationale upon which to construct a genuinely post-liberal social theory. I suggest that deconstructive philosophy might supply such a rationale. This new theory would at the same time be immediately practical, because deconstruction discredits oppositions within liberal Reason such as those of thought and practice, semantics and syntax, mind and body, oppositions that permit a supposedly natural form of knowledge to be dissociated from institutions, history, and technology.

Deconstructive concepts name empirical things, events, and forces, but the implied institutional forms of these concepts, like those of certain of Marx's concepts, cannot be realized as long as liberal Reason reigns in conjunction with liberal social institutions. It is the assumption of deconstruction, as of certain forms of Marxism, that what it conceptualizes exists in latent form within the present system of knowledge or of politics. The Marxist theory of economic crisis suggests that capitalism gives rise to an imbalance of over-production of goods and an under-realization of profit, a simultaneous excess and deficiency that cannot be harmonized within the conceptual-institutional form of liberal capitalism. Over-production is the possibility of communism latent in capitalism, an excess that cannot be controlled and that forces capitalism beyond itself toward communism. Deconstructive theory claims that metaphysical thinking of the sort operative in liberal Reason contains a similarly immanent force that exceeds its liberal conceptual-institutional boundaries and that moves it toward more fallibilist, less absolutist epistemological positions and more democratic, egalitarian, and cooperative socio-institutional forms. There are many names for that "force." Here I shall use the word "dis-placement" for reasons that will become apparent in a moment.

Displacement names a combined process of comparison and distinction, similitude and differentiation. It is the process in rhetoric that permits one thing to be compared or identified with another, by virtue of contiguity or analogy: "My love is like a rose," "The sunshine of your eyes," etc. Such artificial contrivances are seen as secondary and degraded from the point of view of the natural light of Reason, which prefers self-evident ideal or natural foundations. Such contrivances have no natural basis that can serve as an axiomatic foundation for thought. Indeed, rather than provide a ground of authority for the determination of absolute truth, they under-mine such authority. Yet, it can be argued that the supposedly self-evident bases of liberal social theory are products not of the light of liberal Reason but of displacements.

Liberal social theory is founded on displacement because it is based on two founding metaphors or analogies—the metaphor of nature, which connotes freedom and lack of restraint, and the metaphor of scientific law, which suggests order and harmony between parts subsumed under a whole. These two metaphors, when instituted as social practice or as political principles, serve two ideological requirements of the capitalist system of which liberalism is an integral part. The first (nature) sanctifies ownership and the free movements of the entrepreneurial individual in the private accumulation of property. The second (science) provides a justification for the political and juridical forms (political sovereignty, contractual law) that maintain a frame of order around the anarchy of the market and guarantee that although plays may be won or lost, the game will continue.

Liberalism consists of a number of concepts and institutions that form a coherent system only once the initial analogies are accepted. The initial analogies are, first, that there are natural laws that prescribe that everyone in civil society is equal and free; and, second, that civil society functions according to laws that are analogous to the laws of natural science. That is, according to the second analogy, civil society is an orderly and rational machine. From the first analogy derives the doctrine of individual rights. Everyone, as a free, that is, self-possessing individual, has the same right to speech, religion, property, and so on. From the second analogy is derived the doctrine of sovereignty; the state as a whole must be maintained as an integral, orderly machine. Hence, either compromises must be worked out between individual rights and responsibility to the whole (Mill, *On Liberty*), or a theory must be advanced which claims that the free "natural" exercise of individual rights and self-interest results in social harmony (Smith, *The Wealth of Nations*). The institution of political sovereignty mediates between natural right or equality and the actual social inequality caused by the disproportion in property ownership that liberal theory is designed to legitimate. Liberal theory accomplishes this legitimation by recourse either to the concept of a contractual exchange of rights for security (Hobbes, *Leviathan*) or to the concept of a scientific ratio or proportion in civil society between unequal parts (Rousseau, *The Social Contract*).

Not only does liberal civil society assume metaphorical analogies between itself and science and nature, its structural origin is itself a metaphorical process. Metaphor is the rhetorical trope of substitution, displacement, and transference. One thing (a rose) is substituted for another (love) that it displaces (my love is a rose). As a result, the meaning of one thing is displaced onto another. In liberalism, civil society substitutes for nature, whose laws it displaces in favor of civil law. The origin of civil society is always described in liberal theory as a displacement from nature to civil society. In metaphor a comparative equivalence is established between two terms. So in liberal theory, civil society and nature are sometimes seen as being interchangeable.

Thus Thomas Paine: "All the great laws of society are laws of nature."[4] Or Locke, in whose writing civil society is compared by analogy to both nature and science:

> *Freedom of Men under Government* is . . . common to everyone of that society. . . . as *Freedom of Nature* is to be under no other restraint but the law of Nature.[5]

Or Rousseau, in *The Social Contract*:

> Just as nature has set bounds to the stature of a well-formed man . . . , so in what concerns the best constitution of a state, there are limits. . . . Administration becomes more difficult over great distances, just as a weight becomes heavier at the end of a long lever. . . . The state if it is to have strength must give itself a solid foundation . . . for all peoples generate a kind of centrifugal force . . . like the vortices of Descartes.[6]

It is important to point to this constituting role of the structure and movement of metaphor for two reasons. First, it emphasizes the discursive character of liberalism. Liberal institutions are not founded on anything that can be called "real." The institutions of liberal social theory (like property, personhood, universal formal law, and so on) are discursive fictions and constructed metaphors that function ideologically to legitimate class power and rule. For example, nature is both a pre-civil origin and a discursive institution in liberal theory. Nature (defined as "civil war," the prevalence of force over law, and the absence of any guarantee of property right) is the name for what must be displaced if political sovereignty and property are to be guaranteed.[7] In other words, the concept of nature serves a logical, discursive, and ideological function. As a myth of origin, it permits the construction of a genealogical history whose purpose is to make liberal-capitalist forms seem higher and more progressive. Any threat to property right as it is sustained by rational law will thus constitute a regression to nature; nature as historical origin and nature as other to be displaced are inseparable.

Similarly, the structure of metaphor entails the total substitution of one term for another, and liberal civil society is supposed to substitute for nature as a totality. Nature was arbitrary and particularistic; liberal law is necessary and universal. The "natural," uncivil, particularistic force of liberal civility is successfully concealed by this discursive sleight of hand. The particular interest of a few property owners is generalized to the totality of society. That rhetorical maneuver is concrete and real in the institutions, belief systems, and laws of capitalist society, but is no less discursive in character. Indeed, its discursivity is the mark of its ideological character and a sign that the liberal script can be rewritten.

Second, the metaphors at the basis of liberalism emphasize the extent to which it assumes a principle or an operation of displaceability. Civil society is the displacement from nature to civility or law,[8] and that entails transferring rights from the people to the sovereign. As Locke writes:

> Where-ever therefore any number of Men are so united into one Society, as to quit every one his Executive Power of the Law of Nature, and to resign it to the publick, there and there only is a *Political, or Civil Society*. . . . And this *puts Men* out of a State of Nature *into* that of a *Commonwealth*, by setting up a Judge on Earth, with Authority to determine all the Controversies.[9]

Natural rights, therefore, are inherently alienable, and this must be the case if the transfer to civil society is to occur. Natural rights are never fully natural; they are displaceable, that is, contractable and socially exchangeable, from the outset. The anchor of liberalism—natural law—is thus made unstable by precisely that which the anchor supports or to which it gives rise—displacement. The supplement to nature dislocates nature as a self-identical ground or value and sets it loose in a system of displaceability, which is also a system of contract and convention.

Displaceability is born at the same moment as natural right. There is no moment prior to the possibility of displacement when rights are purely natural. A right is power to dispose (over one self or over something else), and the assertion of that power immediately implies the possible threat of its removal. There is no need for right outside of the possibility of transgression. The assertion of right is always the denial of its naturalness, that is, of its permanence or necessity—its non-removability or inalienability. For if rights can be alienated or transferred, they are not "naturally" inalienable. They are constitutively removeable or displaceable. The purpose of liberal theory is to make certain rights, especially property right, non-removeable and non-displaceable, and it accomplishes this task by calling those rights natural. But those rights are only natural through convention, the removal through convention of the possibility of displacement implicit in the very notion of right.

The dehiscence that opens nature to convention and makes arbitrary the seemingly natural attachment of right is nowhere more evident than in that most crucial right of liberal theory, the sanctification of which is the purpose of liberal civil law—property right. The right to own property implies that some other might own it; property right would not need to be assured by law if this were not the case. The self-identity of ownership is constitutively divided by the possibility of transference, the possibility, necessarily inscribed in property right, of someone else's potential ownership of the same thing. The infinite displaceability of property is at once affirmed and denied by property right. Property, as Locke puts it, is "subtracted" from the commons. What is "proper" is such only inasmuch as it is not "common." Property right, therefore, is not something inherent or proper to a person; it is the denial of commonality and displaceability. What seems a natural right is in fact a social force. Moreover, displaceability cannot ever be fully neutralized because it is the condition of possibility of property. Ownership as the appropriation or "subtraction" of an object to oneself from commonality (that is, from "nature," as liberal theorists define it) is itself a displacement. Ownership displaces what is other or common to what is own. Displaceability does not derive from property as an evil that must be avoided through law; rather, it gives rise to property. Ownership is merely a displacement that is recognized by convention as legitimate, and nature, as the word for pre-civility in liberal theory, is merely the name for the nonrecognition of property right and for a resultant infinite displaceability of property, the removal of what one has appropriated as one's own by another or by the "commons."

The concept of nature in liberal theory may thus be seen as an ideological antidote to an irreducible commonality or conventionality of rights and to an infinite displaceability of rights that overrun all the limits of proportion and order liberalism needs to establish if it is to fulfill its function of legitimating the rights of capital. One of the antinomies of liberalism is that it cannot avoid affirming the displaceability of rights, even as it seeks to curtail and limit it.

Nature is the other of civil society, the realm of force where no law guarantees propriety, but it is also the concept that lends authority to the laws of civil society by comparing them to natural science. Nature is to be excluded, yet as natural science it is also the measure of civility. Nature as im-propriety is avoided, but nature as the natural laws of nature is retained. The two aspects of nature are essential. Nature must be abandoned if a civil society in which property is guaranteed by law is to be possible. As Hobbes writes: "Where there is no *own* . . . where there is no commonwealth, there is no propriety." Yet outside nature, a conventional and coercive force that operates as civil law to sanction property right must have an authority akin to the laws of nature as science elucidates them. Hobbes once more: "The precepts by which men are guided to avoid that condition (of 'mere nature' or 'anarchy') are the laws of nature."[10]

The point of the deconstructive critique would be that civil society seeks to exclude uncivility through the scientific rationality of law, yet in that very process acknowledges its own noncivility and inadvertently points toward a form of civil society that would be more egalitarian in character. The exclusion of nature and the analogy from natural science to civil society serve ideological ends. They present the exercise of power by a propertied class over a nonpropertied class as a move from pre-civil anarchy to the scientific rationality of law. Action against property can thus be condemned as uncivil. Even a relatively open liberal thinker like Locke makes the civil sanctification of property prior to the preservation of government. Government may be dissolved when

> the *Legislative acts against the Trust* reposed in them, when they endeavour to invade the Property of the Subject, and to make themselves, or any part of the Community, Masters, or Arbitrary Disposers of the Lives, Liberties, or Fortunes of the People.[11]

What makes civil society civil, makes it simultaneously uncivil—this would be the deconstructive argument in a nutshell. There would be no need to guarantee property by law if nature (as im-propriety or non-ownership) were not a reality within civil society. Civil society contains its other in a state of repression because "nature" is never left behind in the passage to civil society and law. The necessity of law indicates a threat to property from a class of non-owners, a threat to revert from property law to a state of "nature" where property would not be lawful. But "nature" is also present in civil society in one other way, within civil law itself. By its very existence as law, property right testifies to its non-universality and non-uniformity. Its authority resides in the metaphoric analogy to the laws of natural science, which must be uniform and universal. Yet in liberal civil society, property ownership can only be universal as a formal right, not as a law guaranteeing substantive and uniform property ownership itself. Now, this is precisely the characteristic of nature that civil society supposedly excluded and leaves behind. Civil society is "natural" in that it posits a right that cannot be guaranteed by law. Everyone has the right to own property, but no law under liberalism can guarantee that everyone will own property. The necessity of law to guarantee property in civil society indicates that certain people in society can relate to property only as a right, not as a reality. Law, therefore, points out the "natural," that is, particularistic and antagonistic character of liberal civil society.

The necessity of a universal law to guarantee property indicates that property itself can never be as universal in practice as the law that sanctions it in theory. This casts property law more as a force of exclusion or repression (an act of "nature," in other words) than as a rational or scientific principle. The mark of civility is simultaneously a sign of the uncivility of civil society. A progressive possibility derives from this deconstructive antinomy. The scientific principles of universality and uniformity are coercive when imposed on a disproportionate reality. A realization of the principles would imply closing the distance between formal universality and substantive universality. That would entail a move from a universal right to property to a universal equality or uniformity of ownership. And that, of course, would mean the disappearance of property as liberalism conceives it. Liberalism thus provides the instruments of its own dissolution. Liberal civil society, while displaying its own noncivility, points the way toward true civility.

This deconstructive critique of the nature-civil society opposition in liberal theory leads us to a consideration of another implication of the founding metaphors of liberalism—the problem of displacement. The way the scientific principle of universality extends rights beyond the limits placed on them under liberalism constitutes an example of the sort of overrun that is implied by displacement. Displacement is a condition of the possibility of liberalism that is simultaneously a condition of its impossibility as something proper or self-identical. The basis of liberalism is a principle that moves it inevitably beyond its own outlines. Liberalism is founded on a discursive displacement from nature to civil society and a transfer of rights from subjects to a sovereign. But liberalism cannot curtail the transfer of rights beyond its own requirements. Why should rights not be transferred to workers or to women? If government can be transferred initially, why should it not be infinitely transferable? Liberal theorists will seek to find limits for the displacement of rights and of the power over oneself that rights afford, but those limits will simply be the markers of an uncontrollable overrun—closed doors whose overly precipitous closing testify to an anxiety over a very real possibility of opening.

Take the following example from Montesquieu's *The Spirit of Laws*. For Montesquieu there are only three kinds of government, each one with its own particular form of corruption. But the corruption of republican democracy as Montesquieu theorizes it, is not so much an external accident or evil as an indicator of the possibility within civil society of the dissemination of power beyond the bounds set by liberalism and toward a fourth form of government. According to Montesquieu:

> The principle of democracy is corrupted not only when the spirit of equality is extinct, but likewise when they fall into a spirit of extreme equality, and when each citizen would fain be upon a level with those whom he has chosen to command him. Then the people, incapable of bearing the very power they have delegated, want to manage everything themselves. . . . This license will soon become general. . . . Wives, children, slaves will shake off all subjection. No longer will there be any such thing as manners, order, or virtue.[12]

Montesquieu provides an accurate description of what would occur if the founding principles of liberalism were allowed to become indefinitely displaceable, if the displacement that conditions them were not curtailed by law, custom, or force. Equality, rather than being a right to be restrained and limited by law ("In the state of nature, all men are born equal, but they cannot continue in this equality. Society makes them lose it, and they recover it only by the protection of laws.")[13] would become an unrestrainable force, a right to be claimed by all. Like Hobbes and Locke before him, Montesquieu recognizes the danger of displacement. Rights transferred to a sovereign can be transferred back; rights transferred from nature to propertied males can be assumed by "wives, children, slaves."[14]

Fittingly, one of the first liberal theorists to assert the principle of the transferability of rights beyond the bounds set by classical liberal theory was a feminist—Mary Wollstonecraft. In *A Vindication of the Rights of Woman*, she writes: "Let woman share the rights, and she will emulate the virtues of man."[15] Patriarchal hegemony is not merely a secondary characteristic of liberalism, however; in its first theoretical formulations, liberal doctrine is in essence patriarchal. Property right depends on the ability to know who owns what and who inherits what. The transferability of property is controlled by the name of the father passed to the son. Patriarchy is thus crucial to the preservation of property, the very basis of liberal civil society. The patronymic is the sign of the masculinist essence of liberalism. But like other markers of propriety, control, or restraint in liberalism, it testifies to a danger of unlimited displacement, even as it defends against it. Property must be transferred, that is, preserved, if liberal civil society is to survive, but what assures survival also carries the possibility of liberalism's own supersession. The patronymic restrains but simultaneously indicates the reality of the possibly uncontrollable transfer of property.

Wollstonecraft's promotion of the transferability of rights from men to women therefore concerns more than the mere broadening of civic freedoms. It touches the heart of liberalism's commitment to property right and to the political economic power of property-owning males. John Stuart Mill provides liberalism's answer to this challenge by acknowledging women's rights in principle while calling for their non-exercise in practice. Woman is "entitled to choose her pursuits . . . entitled to exert the share of influence on all human concerns which belongs to an individual opinion, whether she attempted actual participation in them or not."[16] The argument follows the outlines of liberalism's justification of property. Property is an abstract, formal right, and as such it is universal. But it can never be universal as a concrete and substantive reality. The form of the universal principle is itself a way of protecting against the effects of the real non-universality of property. Similarly, women are accorded rights in principle, but in practice, Mill argues, their first duties are in the home. In these liberal arguments, a tautology of power is at work. For as the abstract and theoretical (or metapractical) right to property reflects the attitude of abstract and nonlaboring (or metapractical) ownership, so the abstract principle that bestows yet withdraws rights from women is, according to Mill, due to a form of thinking with which men, more than women, are endowed: "A woman seldom runs wild after an abstraction."[17]

Along with feminism, socialism is the other movement that promotes a displacement of rights beyond the bounds established by the founding fathers of liberalism. Rights are first demanded for the poor, then for workers and other dispossessed people. Paine makes social welfare a right: "This support for the poor and the elderly . . . is not of the nature of a charity, but of a right."[18] Marx turns the Lockean principle of property right against liberalism. If labor determines property right, the workers who produce commodities should own them. Like the liberal response to feminism, the response to socialism follows rationalist lines. The formal ,proportion of the market, predicated upon the rational choices of individuals, is rational, while planning is irrational and inefficient.[19] The mathematical rationalism of marginalist theory—an example would be the work of Friedrich Hayek—subsumes questions of social justice into formal problems of proportion.[20] And where the calculus fails, the formal universalist rationality of law (political power) will assure harmony.[21] As a result of the analogy with science as liberal Reason conceives it, rational law ceases to be merely descriptive and becomes prescriptive. The hypothesis of scientific order becomes a social norm. Knowledge, if you will, becomes politics.

Spragens reaches this conclusion. The result of liberal Reason in its absolutist phase is technocracy and in its skeptical phase the irrationalist objectivism of supposedly value-free, purely rational economic and political mechanisms or calculi. Yet these negative results testify to something against which they are reactions, a new epistemology and a new social form emerging out of liberalism. That new epistemology would acknowledge the constitutive role of displacement, the process of comparison and contrast, analogy and differentiation that produces the apparently natural, self-evident, or ideal foundations of liberal Reason. The new social form would be predicated upon the displaceability of the fundamental social categories of liberalism—the individual, the doctrine of rights, property, and political sovereignty.

The deconstructive principle of the undecidable mediation of opposites, without resolution, is the only solution to the antinomies of liberalism and to the twin tendencies of solving those antinomies by absolutizing either subjective rationalism or nonrational objectivism. By deconstructing the oppositions of liberal epistemology, one can begin to formulate the social principles that would be necessary in a post-liberal society in which social institutions would be underwritten by post-liberal Reason. As in Marxist theory, validity is gained for this argument by demonstrating how the deconstructive principle is already active in the liberal world, but in occluded or latent form.

Over-production is the Marxist principle that is curtailed by capitalism because it would create the necessary conditions for that which would abolish capitalism—communism. I have suggested that displacement might be a corollary principle from a deconstructive perspective. Displacement must carry the burden of several meanings here. It implies the displacement of all the binary oppositions I have ascribed to liberalism, oppositions such as subject and object that underwrite both rationalist technocracy and irrationalist objectivism. It also displaces the individual-society opposition in such a way that the doctrine of rights is supplemented by a doctrine of social responsibility. That in turn implies a displacement of fact and value, as well as singular cause and multiple context. The displacement of the individual also displaces the individual mind or logos into an ethical, practical, and collective dimension. Individual reason may determine rational truth, but it does so on the basis of socio-historical bases and contexts, and such truth will have effects on the social context in which it is produced. Truth as theoretical right or authority would be modified to include a practical principle of truth as responsible action.

I will conclude with an example from the liberal tradition of what I have just described as the occlusion of displacement. Darwin describes a world of necessary contextual relations and responsibilities where one part's right is mediated by collective needs and where the doctrine of singular rational cause is supplemented by practical systemic relations and constraints. Yet he analogizes from liberal theory to natural science and thus organizes the evidence of undecidability he describes in nature with an interpretative grid that declares decisively in favor of liberal principles. This is evident in his metaphors of competition, rank, entitlement, individuality, profit, subordination, and so on, as well as in his account of how he arrived at his conclusions: "This is the doctrine of Malthus, applied to the whole animal and vegetable kingdoms."[22]

Darwin describes natural life as a liberal-capitalist economy in which the strong competitors survive. In this way, rank and order are assured; a kind of Adam Smithian harmony results from the general pursuit of self-interest: "All those exquisite adaptations of one part of the organization to another part . . . follow from the struggle for life" (*OS*, pp. 75–76). Darwin chooses to emphasize competition between individuals, but at a number of points in the text his description indicates a certain undecidability between individual competition and structural or systematic inter-relations. What he describes could just as easily be interpreted structurally, relationally, contextually, or differentially as an eco-system in which the different parts necessarily relate to, depend on, and support each other. Yet he consistently chooses the liberal interpretative grid, describing possibly systemic processes in liberal individualist terms:

> Thus I can understand how a flower and a bee might slowly become, either simultaneously or one after the other, modified and adapted to each other in the most perfect manner, by the continued preservation of all the individuals which presented slight deviations of structure mutually favourable to each other. . . . Battle within battle must be continually recurring with

varying success; and yet in the long run the forces are so nicely balanced, that the face of nature remains for long periods of time uniform, though assuredly the merest trifle would give the victory to one organic being over another (OS, pp. 104, 85).

The most telling example of his decisive interpretation of something his own observations seem to indicate is undecidable is the metaphor of the tree that closes the chapter on natural selection:

As buds give rise by growth to fresh buds, and these, if vigorous, branch out and overtop on all sides many a feebler branch, so by generation I believe it has been with the great Tree of Life, which fills with its dead and broken branches the crust of the earth, and covers the surface with its everbranching and beautiful ramifications (OS, p. 137).

Darwin privileges those fresh buds, which, through good fortune and positioning, happen to thrive. By focusing on these stronger and more "competitive" individuals, he plays down the fact that a tree is an aggregational and relational system in which parts lend each other support. The "feebler branch" in the liberal interpretive scheme seems "subordinate," but, from a more systemic or structural point of view, the fresh buds can be said to depend on it.

The example of Darwin is symptomatic of liberal thinking in general in that he provides evidence of undecidability and displacement, yet occludes these in favor of an interpretative grid more favorable (because more analogous) to power relations of liberal capitalism. It is no doubt true that the strong individual survives in a pre-historical, pre-technological, pre-rational setting, but even in such a setting he does so in order to guarantee the survival of the collective. The two principles—individualism and collectivism—are undecidably intertwined.

The principle of infallible, decisively singular truth as absolute cognitive certainty in liberal Reason is intolerant of the undecidability (of individual and collective, universal and particular, fact and value, subject and object, etc.) that is in fact the case in the world. Undecidability is "true," that is existent, but it escapes the rationalist classificatory scheme of liberal Reason, founded as it is on binary oppositions that undecidability shows to be suspect. One symptom of this blindness of liberal Reason is the fact that the very practice of displacement—from nature and science to civil society—that provides the theoretical foundations for liberal theory cannot itself be theorized by liberal Reason. To think the undecidability of individual difference and trans-individual collectivity, for example, requires another, more deconstructive form of Reason. And that Reason would imply a different set of social institutions that would be more democratic, egalitarian, and socialist in character.

NOTES

1. Thomas A. Spragens, Jr., *The Irony of Liberal Reason* (Chicago: University of Chicago Press, 1981).
2. Roberto Mangabeira Unger, *Knowledge and Politics* (New York: The Free Press, 1975).
3. Spragens, *The Irony of Liberal Reason*, pp. 22–23.
4. Thomas Paine, *The Rights of Man* (London: Penguin, 1979), p. 187.
5. John Locke, *Two Treatises of Government* (New York: Cambridge University Press, 1960), p. 324.
6. Jean Jacques Rousseau, *The Social Contract* (London: Penguin, 1979), pp. 90–92.
7. This is clear in Friedrich A. Hayek's *Law, Legislation, & Liberty: The Mirage of Social Justice* (London: Routledge and Kegan Paul, 1976), a recent reassertion of liberal doctrine. Nature, for Hayek, is not a pre-historical category, but rather a name for post-liberal, anti-capitalist socialist experiments: "The demand for 'social justice' is indeed an expression of revolt of the tribal spirit against the abstract requirements of the coherence of the Great Society with no such visible common purpose" (p. 144).
8. See C. B. MacPherson, *The Political Theory of Possessive Individualism* (Oxford: Clarendon Press, 1962). "The agreement to enter civil society does not create any new rights; it simply *transfers* to civil authority the powers men had in the state of nature to protect their natural rights" (p. 218).
9. Locke, *Two Treatises*, pp. 368–69.
10. Thomas Hobbes, *Leviathan* (London: Macmillan, 1962), p. 113, 261.
11. Locke, *Two Treatises*, p. 460.
12. Charles de Montesquieu, *The Spirit of Laws* (New York: Hafner, 1949), p. 109.
13. Montesquieu, *Spirit of Laws*, p. 111.
14. For a recent example of an argument for the transfer of rights, see Bengst Abrahamson and Anders Brostrom, *The Rights of Labor* (Beverly Hills, Calif.: Sage, 1980).
15. Mary Wollstonecraft, *A Vindication of the Rights of Women* (London: Penguin, 1978), p. 319.
16. John Stuart Mill, "An Essay on the Subjection of Women," in *Essays on Sexual Equality*, ed. Alice S. Rossi (Chicago: University of Chicago Press, 1970), p. 222.
17. Mill, *Essays*, p. 142.
18. Paine, *Rights*, p. 265.
19. At the time in which Hayek wrote, this critique of socialism took the form of a critique of planning. Hayek thought competitive individualism was more efficient, using the capitalist criterion of profitability, rather than the socialist one of need satisfaction. See *Collectivist Economic Planning*, ed. Friedrich A. Hayek (London: Routledge and Kegan Paul, 1931) and Hayek, *The Road to Serfdom* (Chicago: University of Chicago Press, 1944).
20. Hayek makes clear the degree to which formal universalism is privileged over (and used as a weapon to constrain and limit) substantive justice in liberal theory in general: "A necessary, and only apparently paradoxical, result of this is that formal equality before the law is in conflict, and in fact incompatible, with any activity of government deliberately aiming at material or substantive equality of different people, and that any policy aiming directly at a substantive ideal of

distributive justice must lead to the destruction of the Rule of Law" (*Road to Serfdom*, p. 79).

21. Once again, Hayek serves as an example. He criticizes socialist planning for exercising control over the economy, but he excuses authoritarian political forms as long as they do not interfere with the economy. Indeed, the implication is that such a political form may be necessary in order for a truly "liberal" economy to operate efficiently. As in Kant, though, that coercion is generally associated with the ideal, formal universality of "law": "What we need . . . is . . . a superior political power which can hold the economic interests in check, and in the conflict between them can truly hold the scales, because it is itself not mixed up in the economic game. The need is for an international political authority which, without power to direct the different people in what they must do, must be able to restrain them from action which will damage others. . . . It is essential that these powers of the international authority should be strictly circumscribed by the Rule of Law" (*Road to Serfdom*, p. 232.)

22. Charles Darwin, *The Origin of Species* (New York: Collier, 1962), p. 27. Hereafter cited in text as *OS;* all succeeding references are indicated parenthetically in the text.

DISPLACEMENT AND THE
DISCOURSE OF WOMAN

***

Gayatri Chakravorty Spivak

WHEN IN *The Philosophy of Right* Hegel writes of the distinction between thought and object, his example is Adam and Eve:

> Since it is in thought that I am first at home *(bei mir)*, I do not penetrate *(durchbohren)* an object until I understand it; it then ceases to stand over against me and I have taken from it its ownness *(das Eigene)*, that it had for itself against me. Just as Adam says to Eve: "Thou art flesh of my flesh and bone of my bone," so mind says: "This is mind of my mind," and the alienness *(Fremdheit* as opposed to *das Eigene;* alterity as opposed to ownness) disappears.[1]

It would be possible to assemble here a collection of "great passages" from literature and philosophy to show how, unobtrusively but crucially, a certain metaphor of woman has produced (rather than merely illustrated) a discourse that we are obliged "historically" to call the discourse of man.[2] Given the accepted charge of the notions of production and constitution, one might reformulate this: the discourse of man is in the metaphor of woman.

I

Jacques Derrida's critique of phallocentrism can be summarized as follows: the patronymic, in spite of all empirical details of the generation gap, keeps the transcendental ego of the dynasty identical in the eye of the law. By virtue of the father's name the son refers to the father. The irreducible importance of the name and the law in this situation makes it quite clear that the question is not merely one of psycho-socio-sexual behavior but of the production and consolidation of reference and meaning. The desire to make one's progeny represent his presence is akin to the desire to make one's words represent the full meaning of one's intention. Hermeneutic, legal, or patrilinear, it is the prerogative of the phallus to declare itself sovereign source.[3] Its causes are also its effects: a social structure—centered on due process and the law (logocentrism); a structure of argument centered on the sovereignty of the engendering self and the determinacy of meaning (phallogocentrism); a structure of the text centered on the phallus as the determining moment (phallocentrism) or signifier. Can Derrida's critique provide us a network of concept-metaphors that does not appropriate or displace the figure of woman? In order to sketch an answer, I will refer not only to Derrida, but to two of Derrida's acknowledged "creditors" in the business of deconstruction, Nietzsche and Freud.[4] I will not refer to *La Carte postale*, my discussion of which is forthcoming.[5]

The deconstructive structure of how woman "is" is contained in a well-known Nietzschean sentence: "Finally—if one loved them . . . what comes of it inevitably? that they 'give themselves,' even when they—give themselves. The female is so artistic."[6] Or: women impersonate themselves as having an orgasm even at the time of orgasm. Within the historical understanding of women as incapable of orgasm, Nietzsche is arguing that impersonation is woman's only sexual pleasure. At the time of the greatest self-possession-cum-ecstasy, the woman is self-possessed enough to organize a self-(re)presentation without an actual presence (of sexual pleasure) to re-present. This is an originary dis-placement. The virulence of Nietzsche's misogyny occludes an unacknowledged envy: a man cannot fake an orgasm. His pen must write or prove impotent.[7]

For the deconstructive philosopher, who suspects that all (phallogocentric) longing for a transcendent truth as the origin or end of semiotic gestures might be "symptomatic," woman's style becomes exemplary, for *his* style remains obliged to depend upon the stylus or stiletto of the phallus. Or, to quote Derrida reading Nietzsche:

> She writes (herself) [or (is) written—*Elle (s')écrit*]. Style amounts to [or returns to *(revient à)*] her. Rather: if style were (as for Freud the penis is "the normal prototype of the fetish") the man, writing would be woman.[8]

A lot is going on here. Through his critique of Nietzsche, Derrida is questioning both the phallus-privileging of a certain Freud as well as the traditional view, so blindly phallocentric that it gives itself out as general, that "the style is the man." Throughout his work, Derrida asks us to notice that *all* human beings are irreducibly displaced although, in a discourse that privileges the center, women alone have been diagnosed as such; correspondingly, he attempts to displace all centrisms, binary oppositions, or centers. It is my suggestion, however, that the woman who is the "model" for deconstructive discourse remains a woman generalized and defined in terms of the faked orgasm and other varieties of denial. To quote Derrida on Nietzsche again:

> She is twice model, in a contradictory fashion, at once lauded and con-
> demned. . . . (First), like writing. . . . But, insofar as she does not believe,
> herself, in truth . . . she is again the model, this time the good model, or
> rather the bad model as good model: she plays dissimulation, ornament,
> lying, art, the artistic philosophy. . . . (*Ep*, p. 66)

At this point the shadow area between Derrida on Nietzsche and Derrida on Derrida begins to waver. "She is a power of affirmation," Derrida
continues. We are reminded of the opening of his essay:

> The circumspect title for this meeting would be
> *the question of style.*
> But woman will be my *subject.*
> It remains to wonder if that comes to the *same*
> (*revient au même*)—or to the *other.*
> The "question of style," as you no doubt have recognized, is a quotation. I
> wanted to indicate that I shall advance nothing here that does not belong to
> the space cleared in the last two years by readings that open a new phase in
> the process of deconstructive, *that is to say affirmative*, interpretation. (*Ep*,
> pp. 34, 36; italics mine)

Quotation in Derrida is a mark of non-self-identity: the defining predication of a woman, whose very name is changeable.[9] " 'Give themselves' "
is thus distinguished from "give themselves" in Nietzsche's description of
woman. The reader will notice the carefully hedged articulation of the
deconstructive philosopher's desire to usurp "the place of displacement":
between the reminder of an appropriate title and the invocation of the
complicity of the same and the other (philosophical themes of great pres-
tige), comes the sentence: "Woman will be my subject." We give the
"subject" its philosophical value of the capital I. In the place of the writer's
"I" will be woman. But, colloquially, "my subject" means "my object."
Thus, even if "le style" (man?) "revient à elle" (returns or amounts to her)
is an affirmation of "ce qui ne revient pas au père" (that which does not
return or amount to the father), the author of *La question du style*—that
displaced text that does not exist, yet does, of course, as *Éperons*—having
stepped into the place of displacement, has displaced the woman-model
doubly as shuttling between the author's subject and object. If, then, the
"deconstructive" is "affirmative" by way of Nietzsche's woman, who is a
"power of affirmation," we are already within the circuit of what I call
double displacement: in order to secure the gesture of taking the woman as
model, the figure of woman must be doubly displaced. For a type case of
double displacement, I turn to "Femininity," a late text of Freud certainly
as well known as the Nietzschean sentence.[10]

II

Freud's displacement of the subject should not be confused with
Freud's notion of displacement *(Verschiebung)* in the dream-work, which is
one of the techniques of the dream-work to transcribe the latent content of
the dream to its manifest content. The displacement of the subject that is
the theme of deconstruction relates rather to the dream-work in general;
for the dream *as a whole* displaces the text of the latent content into the text
of the manifest content. Freud calls this *Entstellung* (literally "displacement"; more usually translated as "distortion").[11]

Freud expanded the notion of the displacement of the dream-work in
general into an account of the working of the psychic apparatus and
thereby put the subject as such in question. One can produce a reading of
the "metapsychological" rather than the therapeutic Freud to show that
this originarily displaced scene of writing is the scene of woman.[12] Let us
consider Freud's description of woman's originary displacement.

"Psycho-analysis does not wish to describe *(nicht beschreiben will)* what
the female *(das Weib)* is . . . but investigates *(untersucht)* how she comes into
being, how the female develops out of the bisexually disposed child" (*F*
XXII, p. 116; *GW* XV, p. 125). The name of this primordial bisexuality is of
course unisex. "We are now obliged to recognize," Freud writes, "that the
little girl is a little man" (*F* XXII, p. 118; *GW* XV, p. 126).

Here is the moment when woman is displaced out of this primordial
masculinity. One of the crucial predications of the place of displacement—"the second task with which a girl's development is burdened"—is
that the girl-child must change the object of her love. For the boy it never
changes. "But in the Oedipus situation the girl's father has become *(ist
geworden)* her love-object." The unchanged object-situation and the fear of
castration allow the boy to "overcome *(überwinden)* the Oedipus complex":

> The girl is driven out of her attachment to her mother through the influence
> of her envy for the penis and she enters the Oedipus situation as though
> into a haven . . . (She) dismantle(s) [*baut ab*] it late and, even so, imperfectly
> [*unvollkommen*]. (*F* XXIII, p. 129; *GW* XV, p. 138)

Through the subject-object topology of the I (ego) and the it (id), Freud
displaces the structure of the psyche itself. The beginning of sexual difference is also given in the language of subject and object. The boy child is
irreducibly and permanently displaced from the mother, the object of his
desire. But the girl-child is doubly displaced. The boy is born as a subject
that desires to copulate with the object. He has the wherewithal to make a
"proper" sentence, where the copula is intention or desire. The sentence
can be

$$S \text{ (subject)} \xrightarrow{\text{desires}} O \text{ (object)}$$

The girl child is born an uncertain role-player—a little man playing a little girl or vice versa. The object she desires is "wrong"—must be changed. Thus it is not only that her sentence must be revised. It is that she did not have the ingredients to put together a proper sentence in the first place. She is originarily written as

$$\text{(masquerading subject)} \xrightarrow{\text{desires (temporarily)}} \cancel{O} \text{ (wrong object)}$$

I have made this analysis simply to suggest that a deconstructive discourse, even as it criticizes phallocentrism or the sovereignty of consciousness (and thus seeks to displace or "feminize" itself according to a certain logic), must displace the figure of the woman twice over. In Nietzsche and in Freud the critique of phallocentrism is not immediately evident, and the double displacement of woman seems all the clearer:

> There is no essence of woman because woman averts and averts herself from herself. . . . For if woman *is* truth, *she* knows there is no truth, that truth has no place and that no one has the truth. She is woman insofar as she does not believe, herself, in truth, therefore in what she is, in what one believes she is, which therefore she is not. (*Ep*, pp. 50, 52)

Here Derrida interprets what I call double displacement into the sign of an abyss. But perhaps the point is that the deconstructive discourse of man (like the phallocentric one) can declare its own displacement (as the phallocentric its placing) by taking the woman as object or figure. When Derrida suggests that Western discourse is caught within the metaphysical or phallogocentric limit, his point is precisely that man can problematize but not fully disown his status as subject. I do, then, indeed find in deconstruction a "feminization" of the practice of philosophy, and I do not regard it as just another example of the masculine use of woman as instrument of self-assertion. I learn from Derrida's critique of phallocentrism— but I must then go somewhere else with it. A male philosopher can deconstruct the discourse of the power of the phallus as "his own mistake." For him, the desire for the "name of woman" comes with the questioning of the "metaphysical familiarity which so naturally relates the *we* of the philosopher to 'we-men,' to the *we* in the horizon of humanity."[13] This is an unusual and courageous enterprise, not shared by Derrida's male followers.[14]

Yet, "we-women" have never been the heroes of philosophy. When it takes the male philosopher hundreds of pages (not to be able) to answer the question "who, me?", we cannot dismiss our double displacement by saying to ourselves: "In the discourse of affirmative deconstruction, 'we' are a 'female element,' which does not signify 'female person.'" Women armed with deconstruction must beware of becoming Athenas, uncontaminated by the womb, sprung in armor from Father's forehead, ruling against Clytemnestra by privileging marriage, the Law that appropriates the woman's body over the claims of that body as Law. To the question: "Where is there a spur so keen as to compel to murder of a mother?" the presumed answer is: "Marriage appointed by fate 'twixt man and woman is mightier than an oath and Justice is its guardian." The official view of reproduction is: "the mother of what is called her child is not its parent, but only the nurse of the newly implanted germ."[15] This role of Athena, "the professional woman," will come up again at the end of the next section.

III

Let us consider briefly the problem of double displacement in Derrida as he substitutes undecidable feminine figurations for the traditional masculine ones and rewrites the primal scene as the scene of writing.

My first example is the graphic of the hymen as it appears in *La double séance*, Derrida's essay on Mallarmé's occasional piece "Mimique."[16]

The hymen is the figure for undecidability and the "general law of the textual effect" (*Dis*, p. 235) for at least two reasons. First, "metaphorically" it is the ritual celebration of the breaking of the vaginal membrane, and "literally" that membrane remains intact even as it opens up into two lips; second, the walls of the passage that houses the hymen are both inside and outside the body. It describes "the more subtle and patient displacement which, with reference to a Platonic or Hegelian idealism, we here call 'Mallarméan' by convention" (*Dis*, p. 235; I have arranged the word-order to fit my sentence). The indefinitely displaced undecidability of the effect of the text (as hymen) is not the transcendent or totalizable ideal of the patronymic chain. Yet, is there not an agenda unwittingly concealed in formulating *virginity* as the property of the sexually undisclosed challenger of the phallus as master of the dialectics of desire? The hymen is of course at once both itself and not-itself, always operated by a calculated dissymmetry rather than a mere contradiction or reconciliation. Yet if the one term of the dissymmetry is virginity, the other term is marriage, legal certification for appropriation in the interest of the passage of property. We cannot avoid remarking that marriage in *La double séance* remains an unquestioned figure of fulfilled indentification (*Dis*, pp. 237–38).

We must applaud Derrida's displacement of the old feminine metaphor of the truth as (of) unveiling: "The hymen is therefore not the truth of unveiling. There is not *aletheia* (truth as unveiling), only a blink of the hymen."[17] Yet desire here must be expressed as man's desire, if only because it is the only discourse handy. The language of a woman's desire does not enter this enclosure:

> the hymen as a protective screen *(écran)*, jewel case *(écrin;* all reminders of writing—*écriture*— and the written—*écrit)* of virginity, virginal wall, most subtle and invisible veil, which, in front of the hysteron, holds itself *between the inside and the outside of the woman, therefore between desire and accomplishment.* (*Dis*, p. 241; italics mine)

Even within this sympathetic scene, the familiar topoi appear. The operation of the hymen is the "outmanoeuvering *(déjouante*—literally 'unplaying') economy of a seduction" (*Dis*, p. 255). We are reminded of Nietzsche as we notice that, in commenting upon the pantomime of a hilarious wife-murder (Pierrot kills Columbine by tickling the soles of her feet) that Mallarmé comments on in *Mimique,* Derrida writes as follows:

> The crime, the orgasm, is doubly mimed. . . . Its author in fact disappears because Pierrot is (plays) also Columbine. . . . The gestures represent nothing that had ever been or could ever become present: nothing before or after the mimodrama, and in the mimodrama, a crime-orgasm that was never committed. . . . (*Dis*, pp. 228, 238–39)

The faked orgasm now takes center stage. The Pierrot of the pantomime "acts" as the woman "is" ("Pierrot is [plays] Columbine") by faking a faked orgasm which is also a faked crime.

Derrida's law of the textual operation—of reading, writing, philosophizing—makes it finally clear that, however denaturalized and non-empirical these sexual images might be, it is the phallus that learns the trick of coming close to faking the orgasm here, rather than the hymen coming into its own as the indefinitely displaced effect of the text. Thus the hymen is doubly displaced. Its "presence" is appropriately deconstructed, and its curious property appropriated to deliver the signature of the philosopher. Hymen or writing "gets ready to receive the seminal jet *(jet;* also throw) of a throw of dice" (*Dis*, p. 317; the last phrase—*un coup de dés*—is of course a reference to Mallarmé's famous poem; but, following Derrida's well-known signature-games, the passage can also read, "the hymen gets ready to receive the seminal J. of a blow of a D"). In terms of the custodianship of meaning, the philosopher no longer wishes to engender sons but recognizes that, at the limit, the text's semes are scattered irretrievably abroad. But, by a double displacement of the vagina, dissemination remains on the ascendant and the hymen remains reactive. It is "dissemination which *affirms* the always already divided generation of meaning" (*Dis*, p. 300). Textual operation is back to position one and fireworks on the lawn with a now "feminized" phallus: "Dissemination in the fold (*repli*—also withdrawal) of hymen" (*Dis*, p. 303).

One of the many projects of *Glas* is to learn the name of the mother.[18] There is an ideological phallocentrism in Freud that works to control some of his most radical breakthroughs. Derrida has traced this phallocentrism in Lacan, who has written in the name of the "truth of Freud."[19] Now in Lacan's gloss on the Oedipus complex, it is through the discovery of the "name of the father" that the son passes the Oedipal scene and is inserted into the symbolic order or the circuit of the signifier. Upon that circuit, the transcendental signifier remains the phallus. Is it possible to undo this phallocentric scenario by staging the efforts of a critic who seeks to discover the name of the *mother?*

Within the argument from double displacement, this might still be a version of Freud's account of the right object-choice: the son's perennial longing for the mother. Whether interpreted this way or not, it remains the undertaking of the right-hand column of *Glas*, where Derrida writes on some writings of Genet. He needs an eccentric occasion to ask the oblique question of the name of the mother: Genet is an illegitimate homosexual son whose name is—if such an expression can be risked—a matronymic.

(This particular concern, the name or status of the mother, remains implicit in the left-hand column of *Glas* as well. Explicitly, Derrida learns to mourn for fathers: his natural father, Hegel, Nietzsche, Freud. Yet the subject-matter is the matter of the family, the place of mother, sister, wife in the Holy Family, in Greek tragedy, in the early writings of Hegel and Marx, in the story of Hegel's own life. Derrida comments repeatedly on the undisclosed homoeroticism of the official discourse of these phallogocentric philosophers—a discourse supported by the relegation of public homosexuals like Jean Genet to criminality.)

I will not attempt an exhaustive description of this search. Let us consider two sentences toward the end of the Genet column:

> I begin to be jealous of his mother who has been able to change her phallus to infinity without being cut up into pieces. Hypothesis Godcome father in himself (*en soi;* without gender differentiation in French) of not being there. (*G*, p. 290b)

The best way to deal with these lines would be to gloss them as mechanically as possible. Derrida has not been able to articulate the name of Genet's mother. The most he has been able to do is a great L made by the

arrangement of the type—"elle" being French for "she"—cradling or being penetrated by a wedge of emptiness.[20] The lines I quote follow almost
immediately.

Derrida is jealous because she can *displace* herself ad infinitum. She has
stolen a march on the false pride of the phallocentric Idea—which can
merely repeat itself self-identically to infinity. She has taken the phallus
out of the circuit of castration, dismemberment, cutting up *(dé-tailler)*.
With her it is not a question of having or not having the phallus. She can
change it, as if she had a collection of dildos or transvestite underwear.
The Genet column of *Glas* has considered a phantasmagoria of such items,
as evoked by Genet in his own texts.

Such a mother—the outcast male homosexual's vision of mother—is
different from the phallic mother of fetishism. If Derrida is re-writing the
text of Freud here by suggesting that the male homosexual is *not* caught in
the fear of castration by regarding the phallus itself as a representation of
what is not there—a theme of self-castration carefully developed in *Glas*—
he must also suggest that the "feminization" of philosophizing for the male
deconstructor might find its most adequate legend in male homosexuality
defined as criminality, and that it cannot speak for the woman.

Such a recognition of the limits of deconstruction is in the admission
that the shape of *Glas*, standing in here for the deconstructive project,
might be a fetish, an object that the subject regards with superstitious
awe. The book is divided into two columns—Hegel on the left, Genet on
the right, and a slit in between. Derrida relates these two pillars with the
fleece in the middle to Freud's reference "to the circumstance that the
inquisitive boy sought out *(gespäht)* the woman's genitals from below, from
her legs up" (*F* xi, p. 155; *GW* xiv, p. 314). It is the classic case of
fetishism, a uniquely shaped object (his bicolumnar book) that will allow
the subject both to be and not to be a man—to have the phallus and yet
accede to dissemination.

And indeed it is in terms of the concept-metaphor of fetishism that
Derrida gives us a capsule history of the fate of dialectics. I can do no more
here than mark a few moments of that "history." Hegel remarks on the
fetishism of the African savage, who must eat the fetishized ancestor
ceremonially. (*Glas* also is an act of mourning for fathers.) Hegel accuses
Kant of a certain fetishism, since Kant sees the Divine Father merely as a
jealous God, and must thus formulate a Categorical Imperative. (Derrida
supplements the accusation by pointing out that, in French at least, the
Categorical Imperative has the same initials as the fetishistic notion—
saving the mother jealously from the father's phallus—of the Immaculate
Conception: IC.)

The negation of the negation (*Aufhebung*, or sublation), at once denying a thing and preserving it on a higher level, Hegel's chief contribution to the morphology of the self-determination of the concept, was itself, Feuerbach suggested, the absolutely positive move. It may be called fetishistic because it allowed Hegel to keep both presence and its representation.

> Marx then exposes Feuerbach's critical movement. . . . The speculative unity, the secular complicity of philosophy and religion—the former being the truth and essence of the latter, the latter the representation of the former. . . . is the process of sublation. (G, p. 226a)

Marx also relates *Aufhebung* to supporting the Christian "desire for maternity *as well as* virginity" (G, p. 228a).[21] The distance between deconstruction's project of displacement and the dialectic's project of sublation may be charted in terms of the son's longing for the mother. "If *Aufhebung* were a Christian mother" (G, p. 225a)—at once marked and unmarked by the phallus—deconstruction looks for a mother who can change her phallus indefinitely and has an outcast homosexual son. Crudely put, a quarrel of sons is not the model for feminist practice.[22]

The project of philosophy, Derrida continues, as each philosopher presents a more correct picture of the way things are, is not merely to locate the fetish in the text of the precursor, but also to de-fetishize philosophy. "If there were no thing—the thing itself par excellence—(in this case the truth of philosophy), the concept of the fetish would lose its invariant kernel. For the fetish is a substitute—of the thing itself" (G, p. 234a; I have modified the order of the sentences to make a summary). Rather than negating the thing itself—that would merely be another way of positing it— deconstruction gives it the undecidability of the fetish. The thing itself becomes its own substitute. Like the faked orgasm, the thing itself its own fake. Yet the fetish, to qualify as fetish, must carry within itself a trace of the thing itself that it replaces. Deconstruction cannot be pure undecidability. "It constitutes an *economy* of the undecidable. . . . It is not dialectical but plays with the dialectic" (G, p. 235a).

Thus *Glas* must end with an erection of the thing, not merely the oscillation of the phallus as fetish. The distance from the dialectic is measured simply by the fact that "the thing is oblique. It *(elle)* already makes an angle with the ground" (G, p. 292b). Its relationship with the ground (of things) has the obliqueness of an originary fetish. The graphic of that angle can be that large L on page 290b. In French, the "it" of the second sentence above is "elle." Cradled in that angle between the fetish and the thing itself is the word *déjà* (already), separated out of the sentence by two commas. *Glas* makes clear that *déjà* is also a bilingual yes *(ja)* to the D (de)—the initial letter of Derrida's own patronymic—in reverse. It is the assent to the self that one must already have given (an assent at best reversed, never fully displaced.) If the project of *La double séance* finally puts the phallus in the hymen, *Glas* is obliged to put the son with the patronymic in the arms of the phallic mother.

"Hypothesis Godcome father in himself of not being there" *(Hypothèse dieuvenue père en soi de n'être pas là)*. This is the mother of whom, simply reversing Kant's position vis-à-vis the jealous father, Derrida begins to become jealous. As the possessor of the fetish, she carries a substitute of the thing itself—that father in himself; yet as the deconstructed fetish she also carries the trace of the thing itself; through not being there she *is*—one presumes, since the verb of being is strategically suppressed in the sentence—the father in himself. Here again that curious displacement—her separation from Athena or Mary. She allows the philosopher to question the concept of being by having no verb of being; she cannot be named. Yet she remains the miraculous hypothesis—"the supposition, i.e., a fact *placed under* a number of facts as their common support and explanation; though in the majority of instances these hypotheses or suppositions better deserve the name of *hypopoiesis* or suffictions."[23]

IV

One must, then, remember *La double séance* and *Glas* as one reads *Éperons*. In the last the project to feminize philosophizing can be understood in the following way. If a man is obliged to perform by means of a single or singular style (stylus, phallus), he can at least attempt a plural style, always try to fake his orgasms, never speak for himself, be forever on the move away from a place that might be locatable as his own. Like the two other pieces, *Éperons* is an exercise in the plural style, a displaced reversal of what Nietzsche would call the "grand style." Ever complicit with his subject-matter, Derrida tries this plural style to comment on the plurality of Nietzsche's style.

As in the case of *Glas*, my method here will be a mechanical decoding. As one attends to the stylistic orchestration in this decoding spirit, one notices among all the subtleties and indirections and ore-packed rifts a set of four triads:

1. *Le voile/tombe*
   L'èrection tombe
   La signature/tombe
   (Ep, pp. 59, 105, 127)
2. He was, he dreaded such a castrated woman
   He was, he dreaded such a castrating woman
   He was, he loved such an affirming woman
   (Ep, p. 100)

3. Perhaps it was cut out *(prelevée)* somewhere
   Perhaps it was heard here or there
   Perhaps it was the sense of a sentence to be written here or there
   (Ep, p. 97)
4. the three final steps of the essay: *un pas encore*
   (yet another step; or, one not yet), P.S., and P.S. II
   (Ep, pp. 135, 138, 140)

Each of these triads stages a self-dislocation and thus connotes heterogeneity. I have repeatedly pointed out that a structural (not natural or biological) description of heterogeneity (not being homogeneously "in place") in intention and signifying convention might be woman:

> The heterogeneity of the text manifests it well. Nietzsche did not give himself the illusion, analyzed it on the contrary, of knowing anything of these effects, called woman, truth, castration, or of the *ontological* (being-related) effects of presence or absence. *(Ep*, p. 94)

The second triad is a summary of what Derrida thinks "The History of an Error" (a chapter in Nietzsche's *The Twilight of the Idols*) reflects. The sentences describe three psychoanalytic "positions," three subject (man)–object (woman) relations. As Derrida explains *(Ep*, p. 96), the first two sentences are reversals, the third a displacement. The displaced "position" sees the woman as "affirming." Deconstruction "affirms" *(Ep*, p. 36). Deconstruction is or affirms the other (woman) after its simple alterity (otherness) has been reversed and displaced.

"How the 'True World' Finally Became a Fable: History of An Error" is Nietzsche's version of what I have called the feminization of philosophizing. In order to prove that the remark "she becomes female" within this chapter can be unpacked into the triad above, Derrida claims that Nietzsche's bitter thoughts are not about woman's essence, but about a historical change in it owing to the ambiguous status given it by Christianity, the ideology of the castrated. "Thus the truth has not always been woman nor is the woman always truth. They both have a history; together they both form a history" *(Ep*, p. 86).

If one attended to the pronominal genders in the chapter, a different story is read. That so meticulous a reader as Derrida does not attend to them is in itself curious. This other story would be the story of the male philosopher's relationship to and birth from woman as such, the story of sexual difference retold.

Since the world—*die Welt*—is feminine in German, the first words to describe the philosopher's relationship to the true world—*die wahre Welt*—describe the child in the womb or at the breast (as if part of the mother's undifferentiated body): "he lives in her, *he is she (Er lebt in ihr, er ist sie)*." Next, this feminine world has become the *idea* (of the true world), and as such, progressing, "she becomes female *(Sie wird Weib)*." This is the first naming of the female as such. Before this she is merely *sie*, the pronominal referent to the true world. Here, at the moment of sexual differentiation, she is desexualized, she becomes neuter; for *Weib* in German is not only contemptuous but neuter in gender. The rest of the chapter is the story of how to abolish "the true world" (Derrida reads the quotation marks as the mark of the woman). In the final paragraph this displaced and neutered woman is indeed abolished—and through a double displacement: both the true world and the apparent one (both the woman and her representation) are abolished. "Highpoint of mankind; INCIPIT ZARATHUSTRA" (*PN*, p. 486; *CM*, p. 75).

As usual in Nietzsche's plural style it is hard to decide if he is endorsing "truth" or "error," or indeed what perspective will allow us to make that distinction. The title of the chapter is subject to that well-recognized Nietzschean reversibility. "The history of an error" could be "the error of a history" just as "Zur Genealogie der Moral" could be "Zur Moral der Genealogie," or expressions like "die Bildung der Begriffe" (the growth of a concept) or "Das Erkennen erfanden" (invented understanding) could in context be read "der Begriff des Bildes" (the concept of the image) or "die Erfindung erkannten" (understood invention).[24] This is the gesture of putting the author's "place" in question.

If Hegel in the *Aufhebung* wishes to conserve both presence (philosophy) and representation (religion),[25] and if, at the end of *Glas*, Derrida wishes to keep a representation (fetish) that substitutes for a presence that is bent, Nietzsche's problematic desire here seems to be to abolish both the woman (the true world) and her representation (the differentiated female—"the true world"). That curious model of philosophizing he cannot practice but leaves rather for a Zarathustra who is merely announced. I have suggested that, paradoxically, when Derrida follows Nietzsche's lead, it results not in an abolishment but in a distanced embracing, of a doubly displaced woman.

If the pronominal charge of the chapter as a whole and especially of *Sie wird Weib* is noted, Zarathustra is seen to become possible through the desexualization of woman as truth or idea. Zarathustra does not speak in this chapter; the author's "place" is once again in question, for he may be no more than a foreshadower; Nietzsche displaces himself even as he doubly displaces the woman.

It is this last move that Derrida presumably describes as "he was, he loved this woman."

I have tried to show and applaud how Derrida seeks to affirm through the (doubly displaced) figure of the woman. I should like, with some trepidation, to suggest that to keep that deconstructive affirmation intact, Derrida must ignore that the third psychoanalytic position in Nietzsche emerges as a violent negation. Whereas in the case of the other three triads in *Éperons* Derrida stages the heterogeneity and displacement in his text, this triad is given as a continuous one. The negation that would mark the heterogeneity between the first two positions and the third is not disclosed. Negation is a mark of the ego's desire to deny heterogeneity or discontinuity. As Freud writes:

> A negative judgment *(Verurteilung)* is the intellectual substitute for repression; the "no" is the hallmark of repression, a certificate of origin—like, let us say, "Made in Germany." With the help of (mediated by; *vermittelt)* the symbol of negation, thinking frees itself from the restrictions of repression and enriches itself with material that is indispensable for its proper functioning *(Leistung)*. (*F* XIX, p. 236; *GW* XIV, p. 12–13)

If the peculiar and uncharacteristic nature of Derrida's protestation of continuity is not noticed, the identification and love for the affirming woman that Derrida finds in Nietzsche can be given a brutal reading. One can then ask: is the displacement-affirmation of deconstruction merely that the man-woman reversal, the scene of "castration," need no longer be seen as a battleground? That "dread" can be turned into "love" by realizing that the clitoridectomized/ hysterectomized neuter woman might just as well be an animal? (The woman whose sexual pleasure is originarily self-(re)presentative in a way *different* from man's might just as well not have a clitoris: the hymen that remains forever (in)violate, upon which the seed is forever spilled afield in dissemination, has no use for the hysteron.) Is this the scene of violence that is called love in the transformation contained within our only perfect triad? If women have always been used as the instrument of male self-deconstruction, is this philosophy's newest twist?

One can then bring forward Derrida's explanation of the third position:

> Woman is recognized, beyond that double negation [the negation of 2 negation in *Aufhebung*], affirmed as affirmative, dissimulative, artistic, dionysiac power. She is not affirmed by man, but affirms herself [or is affirmed—*s'affirme*—herself], *in herself and within man [en elle-même et dans l'homme]*. (*Ep*, 97; italics mine)

Further:

> And in truth, those women feminists so derided by Nietzsche, they are men. Feminism is nothing but the operation of a woman who wishes to

> resemble a man, the dogmatic philosopher, claiming truth, science, objec-
> tivity, that is to say all the virile illusions, and the castration-effect that
> attaches to them. Feminism wishes castration—also of the woman. Loses
> the style. (*Ep*, p. 64)[26]

The scene changes if one notices Derrida's rider to the consequence of
the "indeterminability of castration" if woman is identified with "the question of style" and opposed to "the strict equivalence between the
affirmation and the negation of castration": "To be developed later, perhaps, in terms of the argument of the *athletic support-belt* [*gaine*—an extremely important theme in *Glas*] in Freud's text on fetishism."[27] Perhaps
Derrida speaks from the irretrievably compromised position of a man with
a self-diagnosed fetish (can there be such a thing?) that substitutes nothing
but the trace of a truth (if there could be such a thing). Under the guise of
a description of the problematics of being a feminist woman, he might be
describing the problematics of being a "woman's man." Then we might
remark that, in the lines immediately preceding our passage, where man
and woman are within quotation marks, if the woman is predictably
described as elusive, the man is given the full blast of the critique of
phallocentrism. It is at least within that frame that feminists are derided as
women who wish to resemble men. It is at least possible to read this as a
lament that in the place of phallocentrism a mere hysterocentrism should
be erected:

> The "woman" takes so little interest in truth, she believes in it so little that
> the truth of her proper subject no longer even concerns her. It is the "man"
> who believes that his discourse on woman or truth *concerns*—such is the
> topographical problem that I sketched, which also slipped away, as usual,
> earlier, relating to castration's undecidable contours—the woman. (*Ep*, p.
> 62)

The voice is at least not given as disinterested or Olympian. As much as
at the end of *Glas*, this is a son caught in the desire for the mother, a man
for the woman:

> The questions of art, of style, of truth do not let themselves be dissociated
> from the question of the woman. But the simple formation of this common
> problematic suspends the question "what is woman?" One can no longer
> seek her, no more than one could search for woman's femininity or
> feminine sexuality. At least one cannot find them according to a known
> mode of concept or knowledge, even if one cannot escape looking for them.
> (*Ep*, p. 70).

This might indeed be a bold description of the feminist's problem of

discourse after the critique of the old ways of knowing. To avoid the
problem is to "make a mistake." Yet, with respect, we cannot share in the
mysterious pathos of the longing: for a reason as simple as that the question of woman in general, asked this way, is *their* question, not *ours*.

Perhaps because we have a "different body" the fetish as woman with
changeable phallus is *on her way to* becoming a transcendental signifier in
these texts. As the radically other she does not *really exist*, yet her name
remains one of the important names for displacement, the special mark of
deconstruction. The difference in the woman's body is also that it exists
too much, as the place of evidence, of the law as writing. I am not
referring to the law in general, the Logos as origin, Speech as putative
identity of voice and consciousness, "all the names of the foundation, of
the principle, or of the center (that) have always designated the invariant
of a presence (*eidos*, *archè*, *telos*, *energeia*, *ousia* [essence, existence, substance, subject], *aletheia*, transcendentality, consciousness, God, man, and
so forth."[28] I am speaking in the narrow sense, of the law as the code of
legitimacy and inheritance.

One version of this "simple" law is written on the woman's body as an
historical instrument of reproduction. A woman has no need to "prove"
maternity. The institution of phallocentric law is congruent with the need
to prove paternity and authority, to secure property by transforming the
child into an alienated object named and possessed by the father, and to
secure property by transforming the woman into a mediating instrument
of the production and passage of property.[29] In this narrow but "effective"
and "real" sense, in the body of the woman as mother, the opposition
between displacement and logocentrism might itself be deconstructed.
Not merely as the undecidable crease of the hymen or envied place of the
fetish, but also as the repressed place of production can the woman stand
as a limit to deconstruction.

V

My attitude towards deconstruction can now be summarized: first, deconstruction is illuminating as a critique of phallocentrism; second, it is
convincing as an argument against the founding of a hysterocentric to
counter a phallocentric discourse; third, as a "feminist" practice itself, it is
caught on the other side of sexual difference.[30] At whatever remove of
"différance" (difference/deferment from/of any decidable statement of
the concept of an identity or difference)[31] *sexual difference is thought*, sexual
differential between "man" and "woman" remains irreducible. It is within
the frame of these remarks that I hope the following parable will be read.

Within this frame, let us imagine a woman who is a (straight) deconstructivist of (traditional male) discourse. Let us assume that her position vis-à-vis the material she interprets is "the same" as that of the male deconstructivist. Thinking of the irreducible sexual differential, she might say: in order to have used the discourse of the phallus as a sign of my power, I was obliged to displace myself from what has been defined as my originary displacement by that very discourse and thus (re)-present for myself a place. Should my gesture of deconstructive practice be a third-degree displacement so that, on the other side of the sexual differential, I can "be myself"? Yet, the project of the critique of phallocentrism-logocentrism is an exposure of the ideology of self-possession—"being myself"—in order to grasp the idea—"the thing itself." Should I not have an attitude parallel to the deconstructive philosopher's attitude to the discourse of the phallus towards any discourse of the womb that might get developed thanks to the sexual differential? What about the further problem of creating "purposively" a discourse of the woman to match an official discourse of the man whose strength is that it is often arbitrary and unmotivated? Deconstruction puts into question the "purposive" activities of a sovereign subject.

A certain historical "differential" now begins to suggest itself. Even if all historical taxonomies are open to question, a minimal historical network must be assumed for interpretation, a network that suggests that the phallocentric discourse is the object of deconstruction because of its co-extensivity with the history of Western metaphysics, a history inseparable from political economy and from the property of man as holder of property. *Whatever their historical determination or conceptual allegiance*, the male users of the phallocentric discourse all trace the itinerary of the suppression of the trace. The differential political implications of putting oneself in the position of accomplice-critic with respect to an at best clandestinely determined hysterocentric subtext that is only today becoming "authoritative" in bourgeois feminism, seems to ask for a different program. The collective project of our feminist critic must always be to rewrite the *social* text so that the historical and sexual differentials are operated together. Part of it is to notice that the argument based on the "power" of the faked orgasm, of being-fetish, and hymen, is, all deconstructive cautions taken, "determined" by that very political and social history that is inseparably co-extensive with phallocentric discourse and, in her case, either unrecorded in accessible ways, or recorded in terms of man.[32] Since she has, indeed, learned the lesson of deconstruction, this rewriting of the social text of motherhood cannot be an establishment of new meanings. It can only be to work away at concept-metaphors that deliberately establish and cast wide a different system of "meanings."

If she confines herself to asking the question of woman (what is

woman?), she might merely be attempting to provide an answer to the
honorable male question: what does woman want? She herself still remains the *object* of the question. To reverse the situation would be to ask
the question of woman as a subject: what am I? That would bring back all
the absolutely convincing deconstructive critiques of the sovereign subject.

The gesture that the "historical moment" requires might be to ask the
"question of man" in that special way—what is man that the itinerary of
his desire creates such a text? Not, in other words, simply, what is man?
All the texts in the world are at our disposal, and the question cannot
flounder into the delusions of a pure "what am I?" Yet it restores to us the
position of the questioning *subject* by virtue of the question-effect, a position that the sexual differential has never allowed women à propos of men
in a licit way. This gesture must continue to supplement the collective and
substantive work of "restoring" woman's history and literature.[33] Otherwise the question "what is man's desire?" asked by women from the
peculiar *sub rosa* position of the doubly-displaced subject will continue to
preserve masculinity's business as usual and produce answers that will
describe themselves, with cruel if unselfconscious irony, as "total
womanhood".

As a literary critic she might fabricate strategic "misreadings," rather
than perpetrating variations on "received" or "receivable" readings, especially upon a woman's text. She might, by the superimposition of a suit-
able allegory, draw a reading out of the text that relates it to the historico-
social differential of the body. This move should, of course, be made
scrupulouly explicit. Since deconstruction successfully puts the ideology
of "correct readings" into question, our friend is content with this
thought.[34] Even more content because, since she has never been considered a custodian of truth anyway (only its mysterious figure), this move
seems to possess the virtue of turning that millennial accusation into a
place of strength. To undo the *double* displacement, as it were, and to
operate from displacement as such, if there can be such a thing. To
produce useful and scrupulous fake readings in the place of the passively
active fake orgasm.

VI

In the recently-published "Law of Genre," Derrida/Blanchot (the identities are, as usual, blurred) steps into the Mother/Daughter relationship.[35]
The daughter is the Law *(la loi)*, in French always in the feminine. Instead
of the Law being of the Father, the irreducible madness of the Law-as-
daughter is seductive ("one time she had me touch her knee," writes

Blanchot) of the male mother, who by this means accedes to a "neuter" voice that is "doubly affirmative," "guarding the opportunity of being woman *(garder la chance d'être femme)* or changing sex" (*LG*, pp. 194, 196, 222, 223).

The Law's proffered knee *(genou)* provides the first person with a bisexual "I/we" *(je/nous)*—the rewriting of the "we-men of the total horizon of humanity" that was sought in 1969 in "Ends of Man." All the verbs of seeing and saying at the end of Derrida's essay are given to an "I" within quotation marks. The crucial verb of being lets the *je/nous* enter the story "where I/we are *(ou je/nous somme)."* What this "I," on its knees *(à genous)*, but also as I/we *(je/nous)*, sees is indeed the Law, in sum. But, playing with *somme* ("sum" as well as "are" in the first person plural), the previous sentence would allow the *je/nous* to be in the place of the Law: *la loi en somme.* If the situation of the Law is written into the situation of father-daughter incest, Athena the Law (giving) daughter can be produced, and the phallic mother circumvented; especially if, changing sex, I myself become a mother:

> he wishes to seduce the law to whom he gives birth [there is a hint of incest in this] and . . . he makes the law afraid. . . . The law's female element [which does not signify a female person] has thus always appealed to: me, I, he, we. (*LG*, pp. 225, 198, 197)

The female element does not signify female person. There is no *elle* among the *je, nous,* and *ils* who accede to voice and being at her expense in the last paragraph.

The father-daughter fantasy is not to be found in the companion piece "Living On/ *Border Lines.*"[36] There the narrator operates the hymen "or the alliance *in the language of the other*" (*LO*, p. 77) by "speak(ing) his *mother tongue* as the language of the other" (*LO*, p. 153).

Mixing of the sexes is one of the chief concerns of "The Law of Genre." Before actually beginning his commentary on Blanchot's *La folie du jour,* Derrida reminds us that "in French, the semantic scale of *genre* is much larger and more expansive than in English, and thus always includes within it the gender" (*LG*, p. 221; not in the French version). On the next page we are shown how, by invoking "beautiful creatures" who are "almost always"—but not invariably—"there is no natural or symbolic law, universal law, or law of a genre/gender here"—women, the man accedes to a bizarre quasi-performative:

> In this risky *(aléatoire)* claim that links affirmation almost always to women, beautiful ones, it is then more than probable that, as long as I say yes, yes, I am a woman and beautiful. Grammatical sex (anatomical as well, in any

case, sex submitted to the law of objectivity), the masculine genre (gender) is thus affected by the affirmation through a risky drift that could always make it other. (*LG*, p. 223; pp. 195–96)

Let us remind ourselves that when this sex-change seems to go the other way, from woman to man, the fetishized phallic mother causes a good deal of anguish, as at the end of *Glas*. We might also say, following Derrida in "Limited Inc," that there is "something like a relationship" between such a point of view about becoming a woman at the stroke of a word and the men who legislate and adjudicate against abortion because they believe they can speak for the woman and her body.[37]

The name of the double displacement that allows the double affirmation is now not merely hymen but double invagination, a double turning-inside-out. This creates a space which is larger than the whole of which it is a part, and allows "participation without belonging" (to the female sex?). As in the customary illustrations of the set theory, Venn diagrams, where the men- and the women-sets intersect, there is the left-handed-people set which is larger than either the men-set or the women-set.

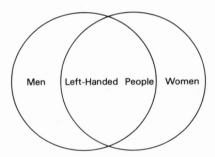

Only here we have fake-frame accounts and fake-interior accounts intersecting (invaginating) to form a "left-handed-people" set where the account-effect can both be and not be a legally answerable account by virtue (if that is the word) of the seductive daughter-law.

(*LG*, p. 218; p. 191)

However stubbornly Derrida might insist that female personhood must be reduced out of the female element or the female silhouette, that the vagina has only a figural connection with invagination, the strength of his own methodology will not allow such a totalizing exclusion and binary opposition to stand. "The opposition of fact and principle . . . in all its metaphysical, ontological, and transcendental forms, has always functioned within the system of what *is*."[38] It is not the question of "a feminist leader" finding it "hard . . . to bear that a 'man' should have dared (such a) 'mad hypothesis' " (*LO*, p. 167). It is rather that other question: what is man that he should, even as he argues deconstruction of the substance-form opposition, need so vehement a negation of woman?

The mathematico-sexual metaphorics of invagination seem even to supersede the self-diagnosed "fetishism" of *Glas*. The telegraphic language allows for great indeterminacy in syntactic connections:

> no piece, no metonymy, no integral corpus. *And thus no fetishism.* Everything said here about double invagination can be brought to bear—a labor of translation—on what is worked out in *Glas*, for example, on the subject of fetishism as the argument of the *gaine* [to be translated "vagina"?]. (*LO*, pp. 137–38; italics mine)

To want to speak for the phallus would, of course, be hopelessly mystified. Fetishism can apparently be circumvented by the morphology of set theory. Perhaps the indefinitely trans-sexual I/we can still not speak for the clitoris as the mark of the sexed subject.

Is it too fanciful to claim that, at the end of "Living On," when Derrida begins the discussion of the "*arrêt* between the two deaths," one can discern a vague legend of the doubly vaginated clitoris (*LO*, p. 163)? "Of course, nothing [or very little] on the manifestly readable surface of the *récits* makes it possible to sustain such a mad hypothesis" (*LO*, p. 170).[39] Yet, it is not the first time Derrida has coupled writing and masturbation:

> *at the very place* where the *relationship* of the "book" *to itself*, in its fragile binding, is formed, the *relationship* of the "I" *to himself*, his alliance with himself, his ring, his anniversary, the *alliance* that joins him to himself. This *very* place, the very *same* place, being the place, the locus, of interruption, is also the place where double invagination gathers together what it interrupts in the strange *same*ness of this place. (*LO*, p. 166)

But perhaps this is circular reasoning. I might see this vague clitoral legend because the space between these two women, two vaginations, or two folds, when the man is not there, has a "terrifying *figura*, figure, face. . . . inter-dicted in the quasi-middle of it, over above beyond its double inner border" (*LO*, p. 166), and is an "uncrossable glass partition" (*LO*, p. 169). Perhaps I am thinking back from passages such as the following and remarking upon the calculable impact of a "different body":

> But the woman touches herself by and in herself without the necessity of a
> mediation, and before all possible decisions between *(départage entre)* activ-
> ity and passivity. A woman "touches herself" all the time, without anyone
> being able to forbid her to do so, in fact, for her sex is composed of two lips
> which embrace continually. Thus, in herself she is already two—but not
> divisible into ones (un[e]s)—who affect each other.
>
> The uncertainty *(suspens)* of this auto-eroticism is effected *(s'opere)* in a
> violent break-in: the brutal spreading apart of this two lips by a violating
> penis.[40]

I haven't a different conclusion to offer. Although, I must repeat, it is a
bold and helpful thing to restore the female element when it is buried in
gender-conventions (I remain surprised that Derrida does not do it in
Éperons), the displacement of the originarily faked orgasm into the mark of
the double affirmation in the interest of man's accession to provisional
androgyny cannot lead us very far. It is excellent to posit this female
element as the irreducible madness of truth-in-law, but we are daily reminded that a little more must be undertaken to budge the law's oppres-
sive sanity. It is not really a question of the "institution" being able to
"bear" our more "apparently revolutionary ideological sorts of 'content'"
(*LO*, p. 95) because we do not threaten its institutionality.[41] It is rather an
awareness that even the strongest personal goodwill on Derrida's part
cannot turn him quite free of the massive enclosure of the male appropriation of woman's voice, with a variety of excuses: this one being, it is not
really woman.

If my present conviction is that to sublate the natural or physiological
evidence of motherhood into a prospective historical or psychological continuity is the idealist subtext of the patriarchal project, what then do I
propose? I have discussed this question at length in "French Feminism in
an International Frame."[42] Here suffice it to indicate the line of my argument somewhat cryptically:

> The clitoris escapes reproductive framing. In legally defining woman as
> object of exchange, passage, or possession in terms of reproduction, it is not
> only the womb that is literally "appropriated"; it is the clitoris as the
> signifier of the sexed subject that is effaced. All historical and theoretical
> investigation into the definition of woman as legal *object*—in or out of
> marriage, or as politico-economic passageway for property and legiti-
> macy—would fall within the investigation of the varieties of the effacement
> of the clitoris.

The social text of motherhood is inscribed within this inquiry. For if an
"at least symbolic clitoridectomy has always been the 'normal' accession to
womanhood and the unacknowledged name of motherhood, why has it
been necessary to plot out the entire geography of female sexuality in

terms of the imagined possibility of the dismemberment of the phallus?"[43]

And when we ask: what is man that the itinerary of his desire creates such a text?—it will help us to remember that the text (of male discourse) gains its coherence by coupling woman with man in a loaded equation and cutting the excess of the clitoris out.

VII

I began this essay with an invocation of great passages. I shall end by going back to the classics as well, and summarize my suggestions as an undoing of the *Eumenides*. I have already written of the immaculate Athena finding against Clytemnestra. Now I speak of another part of her judgment: her defeminating of the Furies, pursuing Orestes the matricide, and bidding them be "sweet-voiced" (Eumenides) by the stroke of a word. If we take the discourse of the "patriarchy" as a straw monster, and pursue it mightily, our role as Furies will lead to little more than self-congratulation and euphoria. We must use and attend to "the patriarchy's" own self-critique even as we recognize that it is irreducibly determined to disable us. It was after all a man who pointed out that the real charge in Hegel's picture of the subject appropriating the object—sweetly metaphorized by Adam and Eve—was a deep hostility:

> The appropriation *(die Aneignung)* of estranged objective being on the sublation *(die Aufhebung)* of objectivity in the determination *(Bestimmung)* of *estrangement*—which must proceed from indifferent alienness *(Fremdheit)* to real hostile estrangement—has for Hegel at the same time or even principally the significance of the sublation *(aufzuheben)* of *objectivity*, since it is not the *determinate (bestimmte)* character of the object but its *objective* character which constitutes the offense and the estrangement for self-consciousness. The object is therefore negative, self-sublating *(sich selbst Aufhebendes)*, a nullity.[44]

Although Derrida and deconstruction, because of their overt critique of phallocentrism, rather than Marx and materialism, have been my example, the entire business of my essay might still be summed up a) in the suggestion that a feminist reader would see in Marx's correction of Hegel a gesture useful for feminism, and b) in the definition of such a reader, such uses.

NOTES

1. Georg Wilhelm Friedrich Hegel, *Sämtliche Werke*, vii (Leipzig: F. Meiner, 1920–55), p. 47; Hegel, *Philosophy of Right*, trans. T. M. Knox (Oxford: Claredon Press, 1942), p. 226. Throughout the essay I have modified all quotations from texts in translation when necessary.

2. I do not use the word "patriarchy"—the rule of the father—because it is susceptible to biologistic, naturalistic, and/or positivist-historical interpretations, and most often provides us with no more (and no less) than a place of accusation. I am more interested in the workings of a certain "discourse"—language in an operative and abyssal heterogeneity. I should add that the absence of Marxist issues in this paper signifies nothing that cannot be explained by the following conviction: as women claim legitimation as agents in a society, a congruent movement to redistribute the forces of production and reproduction in that society must also be undertaken. Otherwise we are reduced to the prevailing philosophy of liberal feminism: "a moralistically humanitarian and egalitarian philosophy of social improvement through the re-education of psychological attitudes" (Charnie Guettel, *Marxism and Feminism* [Toronto: Women's Press, 1974], p. 3). As a deconstructivist, my topic in the present essay is—can deconstruction help? That should not imply that I am blind to the larger issues outlined here.

3. For literary critics, the most recent articulation of this "official philosophy" is in the concept of the hermeneutic circle. Digests can be found in Sarah N. Lawall, *Critics of Consciousness: The Existential Structures of Literature* (Cambridge: Harvard University Press, 1968); and Robert R. Magliola, *Phenomenology and Literature: An Introduction* (West Lafayette: Purdue University Press, 1977).

4. See Jacques Derrida, "Speculations on 'Freud,'" trans. Ian McLeod, *Oxford Literary Review* 3 (1978):78–97.

5. Spivak, "Love Me, Love My Ombre, Elle," forthcoming in *Diacritics*.

6. Friedrich Wilhelm Nietzsche, *Werke; kritische Gesamtausgabe*, v, vol. 2, ed. Georgio Colli and Mazzino Montinari (Berlin: W. De Gruyter, 1970), p. 291, hereafter cited in the text as *CM*; Nietzsche, *The Gay Science*, trans. Walter J. Kaufmann (New York: Vintage Books, 1974), p. 317.

7. I do not believe Nietzsche's passage is necessarily read this way by everyone.

8. Jacques Derrida, *Éperons: Les Styles de Nietzsche; Spurs: Nietzsche's Styles*, trans. Barbara Harlow (Chicago: University of Chicago Press, 1979), p. 56, hereafter cited in the text as *Ep*. This is a bilingual edition of *Éperons;* I have used my own translations.

9. For a discussion of "citationality," see Jacques Derrida, "Limited Inc," trans. Samuel Weber, *Glyph* 2 (1977):162–254. For a discussion of citationality in Derrida, see Spivak, "Revolutions That As Yet Have No Model: Derrida's *Limited Inc*," *Diacritics* 10 (Winter 1980):29–49.

10. Sigmund Freud, *Standard Edition of the Complete Psychological Works*, trans. James Strachey, xxii (London: Hogarth Press, 1964), hereafter cited in the text as *F; Gesammelte Werke*, xv (Frankfurt am Main: S. Fischer, 1940), hereafter cited in the text as *GW*. Citations indicate volume and page number.

11. For definitions of psychoanalytic terms, consult Jean Laplanche and J.-B. Pontalis, *Le Vocabulaire de la psychanalyse* (Paris: Presses Universitaires de France, 1967); *The Language of Psycho-Analysis*, trans. Donald Nicholson-Smith (New York: Norton, 1973). For a cautionary viewpoint against such a sourcebook, see Derrida, "Moi-la psychanalyse," introduction to Nicolas Abraham, *L'Écorce et le noyau* (Paris: Aubier-Montaigne, 1978); "Me-Psychoanalysis: An Introduction to the Translation of *The Shell and the Kernel* by Nicolas Abraham," trans. Richard Klein, *Diacritics* 9 (March 1979):4–12.

12. Derrida produces such a reading, using *Beyond The Pleasure Principle* as his occasion, in "Speculer-sur 'Freud,'" in *La Carte postale* (Paris: Aubier-Flammarion, 1980), pp. 237–437. The full French text has not yet been translated.

13. Jacques Derrida, "Les fins de l'homme," *Marges de la philosophie* (Paris: Minuit, 1972), p. 137; "Ends of Man," trans. Edouard Morot-Sir et al., *Philosophy and Phenomenological Research* 30 (Setpember 1969):35.

14. Since I wrote this essay, Michael Ryan and Jonathan Culler have published studies of deconstruction that include chapters on feminism. See Ryan, *Marxism and Deconstruction: A Critical Articulation* (Baltimore: Johns Hopkins University Press, 1982), pp. 194–212; and Culler, *On Deconstruction: Theory and Criticism after Structuralism* (Ithaca: Cornell University Press, 1982).

15. *Aeschylus*, trans. Herbert W. Smyth, II (London: W. Heinemann, 1936), pp. 311, 293, 335.

16. Jacques Derrida, "La double séance," *La dissémination* (Paris: Seuil, 1972); *Dissemination*, trans. Barbara Johnson (Chicago: University of Chicago Press, 1981), hereafter cited in the text as *Dis*. Page references are to the French edition, and the translations are my own.

17. *Dis*, p. 293. The hymen is here also substituted for the imperious eye, whose blink measures the self-evident moment (in German *Augenblick*, literally the blink of an eye), in Husserlian philosophy as in the general Western tradition; see "Le Signe et le clin d'oeil," *La Voix et le phénomène* (Paris: Presses Universitaires de France, 1967); "Signs and the Blink of an Eye," *Speech and Phenomena*, trans. David Allison (Evanston: Northwestern University Press, 1973).

18. Jacques Derrida, *Glas* (Paris: Galilée, 1974), hereafter cited in the text as *G*.

19. Jacques Derrida, "Le Facteur de la vérité," *Poétique* 21 (1975):96–147; "The Purveyor of Truth," trans. Willis Domingo et al., *Yale French Studies* 52 (1975):31–114.

20. This particular reading of the capital L has been independently developed by Geoffrey Hartman in *Saving the Text* (Baltimore: Johns Hopkins University Press, 1981), p. 75.

21. In "Freud and the Scene of Writing," in *Writing and Difference*, and in *La double séance*, Derrida suggests that both in Freud and in Mallarmé the desire is to find a surface both marked and virgin. In *De la grammatologie* (Paris: Minuit, 1967); *Of Grammatology*, trans. Gayatri Spivak (Baltimore: Johns Hopkins University Press, 1976), he suggests that Rousseau wanted a category that was both transcendental (virgin) and supplementary (marked). An interpretation of Derrida's interpretation of the intellectual history of European men, in terms precisely of sons' longing for mothers, can perhaps be made.

22. I should make it clear that Derrida himself, like the Nietzsche of *Ecce Homo*, is, at least in theory, suspicious of discipleship. This particular "feminist" charge would probably seem a mark of excellence to him.

23. Samuel Taylor Coleridge, *Biographia Literaria*, ed. J. Shawcross, I (London: Oxford University Press, 1907), p. 72.

24. The last two examples are from "Über Wahrheit und Lüge im äusser-moralischen Sinne," *CM* III. 2, Berlin, 1973, p. 373, 369; "Of Truth and Falsity in An Extramoral Sense," *Essays on Metaphor*, ed. Warren Shibles (Whitewater, Wisc.: Language Press 1972), p. 41.

25. This could be related to the idea that Marx saw in Hegel "a double inversion" (first of subject and predicate, and next of idealism and an unexamined empiricism), which Lucio Colletti develops in *Marxism and Hegel* (London: New Left Books, 1973).

26. "The style" of course continues to allude to the phallus, whose status in Derrida is precarious precisely because castration cannot be taken as the all-or-

nothing threat, and the question of the style (phallus) remains "the question of woman."

27. For elaboration upon Derrida's argument from the *gaine*, see my "*Glas*-piece: A Compte Rendu," *Diacritics* 7 (1977):22–43.

28. Jacques Derrida, "La Structure, le signe et le jeu," *Écriture et la différence*, p. 411; "Structure, Sign, and Play," *The Structuralist Controversy: The Languages of Criticism and the Sciences of Man*, ed. Richard Macksey and Eugenio Donato (Baltimore: Johns Hopkins University Press, 1970), p. 249.

29. Although Engels' strict progressivist-dialectical account of the stages of marriage with matching sexual relations of production would be indefinitely complicated by a deconstructivist analysis, his pioneering statement is worth quoting here: "The first class antagonism which appears in history coincides with the development of the antagonism between man and woman in monogamian marriage, and the first class opposition with that of the female sex by the male." In Friedrich Engels, *The Origin of the Family, Private Property, and the State* (New York: Pathfinder Press, 1972), p. 75. The distinction between patrilineage as passage of property and so-called matrilineage would also be open to a deconstructive reading. It should also be remembered that much of Engels' work in this book owes an unacknowledged debt to Flora Tristan.

30. From this point of view, it is worth noting that in Julia Kristeva's more mainstream or masculist celebration of motherhood, the child remains male ("Héréthique de l'amour," *Tel Quel* 74 (Winter 1977):30–49. "Maternité selon Giovanni Bellini" and "Noms de lieu," in *Polylogue* (Paris: Seuil, 1977); "Motherhood According to Bellini" and "Place Names," in *Desire in Language: A Semiotic Approach to Literature and Art*, trans. Thomas Gorz et al. (New York: Columbia University Press, 1980).

31. Derrida, "La Différance," in *Marges;* "Differance," *Speech and Phenomena*.

32. I have tried to develop such a program since this essay was written. See especially Spivak, "Feminism and the Critical Tradition," forthcoming in a collection of essays edited by Paula Treichler, to be published by the University of Illinois Press.

33. Eleanor Fox-Genovese has written a pathbreaking essay on the subject that appeared after my own essay was completed. See her "Placing Women's History in History," *New Left Review* 133 (May–June 1982).

34. I have attempted to use this method of criticism in "Unmaking and Making in *To the Lighthouse*," in *Women and Language in Literature and Society*, ed. Sally McConnell-Ginet and Nelly Furman (New York: Praeger, 1980).

35. Jacques Derrida, "The Law of Genre," tr. Avital Ronell, *Glyph* 7 (1980). Both English and French versions appear in *Glyph;* hereafter cited in the text as *LG* with page reference to the English version followed by reference to the French.

36. Jacques Derrida, "Living On: Border Lines," trans. James Hulbert, in *Deconstruction and Criticism*, ed. Harold Bloom et al. (New York: Seabury Press, 1979), hereafter cited in the text as *LO*.

37. In "Limited Inc," *Glyph* 2 (1977), Derrida points to the relationship between the normativity of speech act theory and the repressive State apparatus of official psychiatry.

38. Derrida, *Grammatologie*, p. 110; *Grammatology*, p. 75.

39. My apologies to David Carroll, who has quite appropriately chastised me for my tendency to imitate Derridian gestures: "Spivak's attempt (and failure) to imitate Derrida's style is precisely the problem with her very long 'Translator's

Preface' to *Of Grammatology*, which I would advise any reader not totally familiar with Derrida's writing simply to ignore." Cf. "History As Writing," *Clio* 7 (Spring 1978): 460.

40. Lucy Irigaray, *Ce sexe qui n'en est pas un* (Paris: Minuit, 1977), p. 24; tr. in *New French Feminisms*, ed. Elaine Marks and Isabelle de Courtivron (Amherst: University of Massachusetts Press, 1980), p. 100.

41. Roland Barthes brought a comparable charge against "'a group of revolutionary students'" in "Écrivains, intellectuels, professeurs," *Tel Quel* 47 (Autumn 1971): 8; tr. Stephen Heath, "Writers, Intellectuals, Teachers," in *Image/Music/-Text* (New York: Hill and Wang, 1977), pp. 198–99. Given the tradition of academic radicalism in France, and our experiences with the old New Left, "feminist" should not be taken as a subset of "revolutionary." For the institutional recuperation of feminist criticism, see Spivak, "A Response to Annette Kolodny," forthcoming in *Signs*.

42. Spivak, "French Feminism in an International Frame," *Yale French Studies* 62 (1981): 154–84.

43. This passage and the preceding one are modified quotations from "French Feminism in an International Frame."

44. Karl Marx and Friedrich Engels, *Werke*, Berlin (1960–68). Ergänzungsband, T.1, p. 579–80; *Early Writings*, tr. Rodney Livingstone and Gregor Benton, (New York:Random House), p. 391.

Contributors

Tom Conley teaches French at the University of Minnesota. His essays on literature, painting, and theory have appeared in journals in America and France, and he is completing one book on Montaigne's essays and another on film.

Paul de Man is Sterling Professor of the Humanities at Yale and the author of *Blindness and Insight* and *Allegories of Reading*. He is preparing a series of essays that will be included in a book about the relationship between rhetorical, aesthetic, and ideological discourse in the period from Kant to Kierkegaard and Marx.

Susan Handelman teaches in the Departments of English and Jewish Studies at the University of Maryland–College Park. She is the author of *The Slayers of Moses: The Emergence of Rabbinic Interpretation in Modern Literary Theory*.

Andrew Parker teaches in the department of English at Amherst College. His essays have appeared in *Boundary II*, *Diacritics*, and *MLN*, and he is completing a manuscript entitled *Re-Marx: Literature, History, and Politics After Derrida*.

Herman Rapaport has recently been a Butler Fellow at SUNY–Buffalo and is currently teaching in the English Department at Loyola University of Chicago. He is the author of *Milton and the Postmodern* and has published a number of articles, mainly in the field of psychoanalysis and literature.

Michael Ryan teaches English at Miami University (Ohio). He has written many articles, chiefly on the politics of the new French theory, and a book, *Marxism and Deconstruction: A Critical Articulation*.

Susan Shapiro teaches in the Religion Department at Syracuse University. She received her doctorate in the Ideas and Methods program of the University of Chicago, with a dissertation on Jewish sacred and quasi-sacred texts from Maimonides to Edmond Jabès.

Gayatri Chakravorty Spivak is professor of English at the University of Texas–Austin and the translator of Derrida's *De la Grammatologie*. She has published widely in the areas of deconstructive and Marxist-feminist criticism, and she is now at work on a book on Feminism and the Social Text incorporating these approaches.

GREGORY ULMER teaches in the English Department at the University of Florida. His major recent project, soon to appear as a book, is entitled *Applied Grammatology: Post(e)-Pedagogy from Jacques Derrida to Joseph Beuys*.

MARK KRUPNICK teaches English at the University of Wisconsin-Milwaukee. His essays, mainly on American criticism, have appeared in *Salmagundi* and other journals. He is completing a book on Lionel Trilling and his New York intellectual contemporaries.